SpringerBriefs in Earth Sciences

SpringerBriefs in Earth Sciences present concise summaries of cutting-edge research and practical applications in all research areas across earth sciences. It publishes peer-reviewed monographs under the editorial supervision of an international advisory board with the aim to publish 8 to 12 weeks after acceptance. Featuring compact volumes of 50 to 125 pages (approx. 20,000–70,000 words), the series covers a range of content from professional to academic such as:

- timely reports of state-of-the art analytical techniques
- bridges between new research results
- snapshots of hot and/or emerging topics
- literature reviews
- in-depth case studies

Briefs will be published as part of Springer's eBook collection, with millions of users worldwide. In addition, Briefs will be available for individual print and electronic purchase. Briefs are characterized by fast, global electronic dissemination, standard publishing contracts, easy-to-use manuscript preparation and formatting guidelines, and expedited production schedules.

Both solicited and unsolicited manuscripts are considered for publication in this series.

More information about this series at http://www.springer.com/series/8897

Marco Keersemaker

Suriname Revisited: Economic Potential of its Mineral Resources

 Springer

Marco Keersemaker
Faculty of Civil Engineering
and Geosciences
Delft University of Technology
Delft, The Netherlands

ISSN 2191-5369   ISSN 2191-5377   (electronic)
SpringerBriefs in Earth Sciences
ISBN 978-3-030-40267-9   ISBN 978-3-030-40268-6   (eBook)
https://doi.org/10.1007/978-3-030-40268-6

This Springer imprint is published by the registered company Springer Nature Switzerland AG
The registered company address is: Gewerbestrasse 11, 6330 Cham, Switzerland

A no ala sani gowtu san brinki
Sranan odo

Preface

It has been an eventful start of the century for Suriname's mining industry: we have witnessed the closure of its bauxite and alumina industry as well as the development of two world-class gold mines and the recent kick-off of the South American Exploration Initiative or SAXI, which will hopefully lead to more developments. As one of the last employees of BHP Billiton Maatschappij Suriname, I was closely involved in the departure of my employer after another failed attempt to develop the Bakhuis bauxite project. That was a painful experience for me and my colleagues.

During my time in the country, I occasionally visited the archives of GMD and BHP Billiton archives and saw numerous publications and maps on the mineral exploration work done in the past. Local and foreign scientists spent many years exploring the interior and identified several mineral occurrences of a range of commodities. Some of these documents from the period after the Second World War are still the most recent reports available and I often wondered if there was a chance that under current market circumstances, any of them would possibly be economic.

Besides my years in Suriname, I have also spent almost a decade working on mines and mine projects (gold, manganese, bauxite and iron ore) in West Africa, which shares a geological history with the Guianas. The difference in mineral endowment and economic mining activities is significant, which strengthened my belief that there must be hidden opportunities in Suriname.

Professor Wong et al. were the last to review in 1998, but the recently commenced South-American Exploration Initiative (SAXI) is putting the spotlight back on Suriname's geology and I considered it an appropriate time to have another look at historic research by experienced and knowledgeable geoscientists and to give my updated interpretation as a mining engineer with an economic perspective in the hope that the country can benefit from our combined contributions.

Delft, The Netherlands   Marco Keersemaker

Acknowledgements

The author would like to thank Dr. Mike Buxton, Prof. Salomon Kroonenberg and John Sew A Tjon for reviewing the manuscript and for their contributions, which helped to improve the quality.

Further thanks go to:

- Prof. Jan Holtrop who donated his collection of reports from his years as researcher in Suriname and who inspired me to make good use of it.
- Dr. Leo Kriegsman and Arike Gill of Naturalis Biodiversity Center for assisting with identification of minerals and allowing access to their mineral archive.
- Dr. Emond de Roever for information on Bakhuis.
- Elsevier and Cambridge University Press for granting permission to use images.
- Janice Rossen for her thorough editing of the draft text and last but not least:
- Ingeborg Ligtenberg for her support and positive energy.

Contents

Abbreviations

AA	Available Alumina
AISC	All-in Sustaining Cost
API	Alumina Pricing Index
BMS	BHP Billiton Maatschappij Suriname
GMD	Geologische Mijnbouwkundige Dienst
LOI	Loss on Ignition
Mtpa	Million tonne per annum
PGM	Platinum Group Metals
REE	Rare Earth Elements
RGM	Rosebel Gold Mines
SAXI	South American Exploration Initiative
USGS	United States Geological Survey
WAXI	West African Exploration Initiative

Chapter 1
Introduction

This book describes the geology and known occurrences of metallic minerals in the Republic of Suriname with a view to assessing their current economic potential.

Suriname is located on the northern Atlantic coast of South America, between 2° and 6° N latitude, and between 54° and 58° W longitude, sharing borders with Guyana, French Guyana, and Brazil. Its population is 573,000 (2018 estimate) about half of which is living in its capital Paramaribo. The greater part of the country is virtually uninhabited, without any roads, and is covered by dense tropical rainforest. As shown in Figs. 7.1 and 7.2, road infrastructure is limited to the coastal areas connecting east and west, with a few links in southern direction towards Pokigron south of Brokopondo Lake, Apoera in the west, and Nassau in the east; most of the interior is accessible by small aircraft and small boat only. The climate is tropical, with an annual mean temperature of 27 °C and rainfall of 2,200 mm.

So far, only gold and bauxite have been mined on a large scale, besides the production of oil. No major deposits of metallic and non-metallic minerals have been found, much less developed, despite the presence of geological occurrences of minerals identified in the past. This could be due to low mineral endowment, as indicated by various studies, but it is also plausible that the challenging environment of dense rain forest, absence or poor condition of infrastructure, and thick lateritic overburden are prohibitive, as they make exploration difficult and expensive. The difference in levels of exploitation between the Marowijne and the Birimian Greenstone belts in West Africa are a case in point, and this comparison will be elaborated further.

The question arises: what we should be looking for, in terms of resource grade and tonnage, to make economic exploitation of the various known minerals feasible? This book attempts to answer that question, in relation to a range of metallic and non-metallic minerals which have been identified as being present in Suriname. (Hydrocarbons are not included in this publication.)

Similar publications have appeared in the past, notably those by Doeve in 1966, Bosma, Ho Len Fat and Welter in 1972, and de Vletter in 1984 who all looked at the subject of Suriname's mineral potential, and whose work has served as valuable references throughout this book. Professor Wong et al. published a book on Suriname's

© The Author(s), under exclusive license to Springer Nature Switzerland AG 2020
M. Keersemaker, *Suriname Revisited: Economic Potential of its Mineral Resources*,
SpringerBriefs in Earth Sciences, https://doi.org/10.1007/978-3-030-40268-6_1

geology in 1998, in which several authors provided updates on earlier publications, and which has been a valuable reference. For more detail on geological context, the reader is referred to publications by specialists, as indicated in the text.

Markets and prices have changed significantly, in the past 20 years since the last publication on Suriname's mineral potential. Demand driven by China's growth, urbanization trends and the new economy (based on information technology and energy transition) has pushed prices to new levels for a range of commodities. Conditions may have changed, to now make previously uneconomic deposits targets for feasible exploitation, or at least to make them subjects of renewed efforts to study their potential. The following chapters will address those issues in relation to a range of relevant minerals.

The mineral occurrences identified in Suriname are discussed in individual paragraphs, in the course of this book. Whether any of these occurrences have the potential to be exploited at positive return of investment remains to be determined. This book attempts to provide the reader with ballpark estimates of the minimum deposit parameters: tonnage and grade at current market conditions, some of which have changed significantly since previous publications on this subject.

Chapter 2
Geology of Suriname

About 80% of Suriname is occupied by a Precambrian crystalline shield, which forms part of the Guiana Shield and consists largely of granitoids and associated metavolcanics, in which three distinct belts of regionally metamorphosed belts occur. Near the center of the country, the basement is locally overlain by the subhorizontal Middle Proterozoic Roraima sandstone Formation. All these rocks are transected by Precambrian and Permo-Triassic dolerite dikes with predominantly north-eastern and north-western directions (IJzerman 1931).

In the coastal area to the north, the basement is covered by sediments from late Cretaceous to recent times. See Fig. 2.1, with an updated, simplified geological map of Suriname.

The crystalline basement was formed essentially in the Lower Proterozoic during the Trans-Amazonian Orogenic Cycle. The Trans-Amazonian Orogeny took place between 2.26–1.98 billion years ago and affected the whole northern and central part of the Guiana Shield, eventually resulting in the collision of the Amazonian and West African Cratons (Kroonenberg et al. 2019). The above-mentioned metamorphic belts consist of:

- The Falawatra Group granulite belt in the northeast-trending Bakhuis Horst in northwest Suriname. The Group consists mainly of banded charnockitic granulites.
- The amphibolite-granulite facies Coeroeni Group, largely made up of quartzofelds-pathic and pelitic gneisses (Kroonenberg 1976).
- The low—to medium-grade metavolcanic-sedimentary greenstone belt (Marow-ijne Group) in northern and northeast Suriname, which is of major economic potential (Bosma et al. 1984).

Most mineralisations occurred in a period of rapid crustal growth, during the formation of the Atlantic paleocontinent at 2.1–2.0 billion years ago (Texeira et al. 2007; Goldfarb et al. 2017), which eventually became a part of the Columbia super-continent from—1.8 billion years. Extensive acid magmatism marked the second stage in the Trans-Amazonian Orogenic Cycle, resulting in the outpouring of large

© The Author(s), under exclusive license to Springer Nature Switzerland AG 2020
M. Keersemaker, *Suriname Revisited: Economic Potential of its Mineral Resources*,
SpringerBriefs in Earth Sciences, https://doi.org/10.1007/978-3-030-40268-6_2

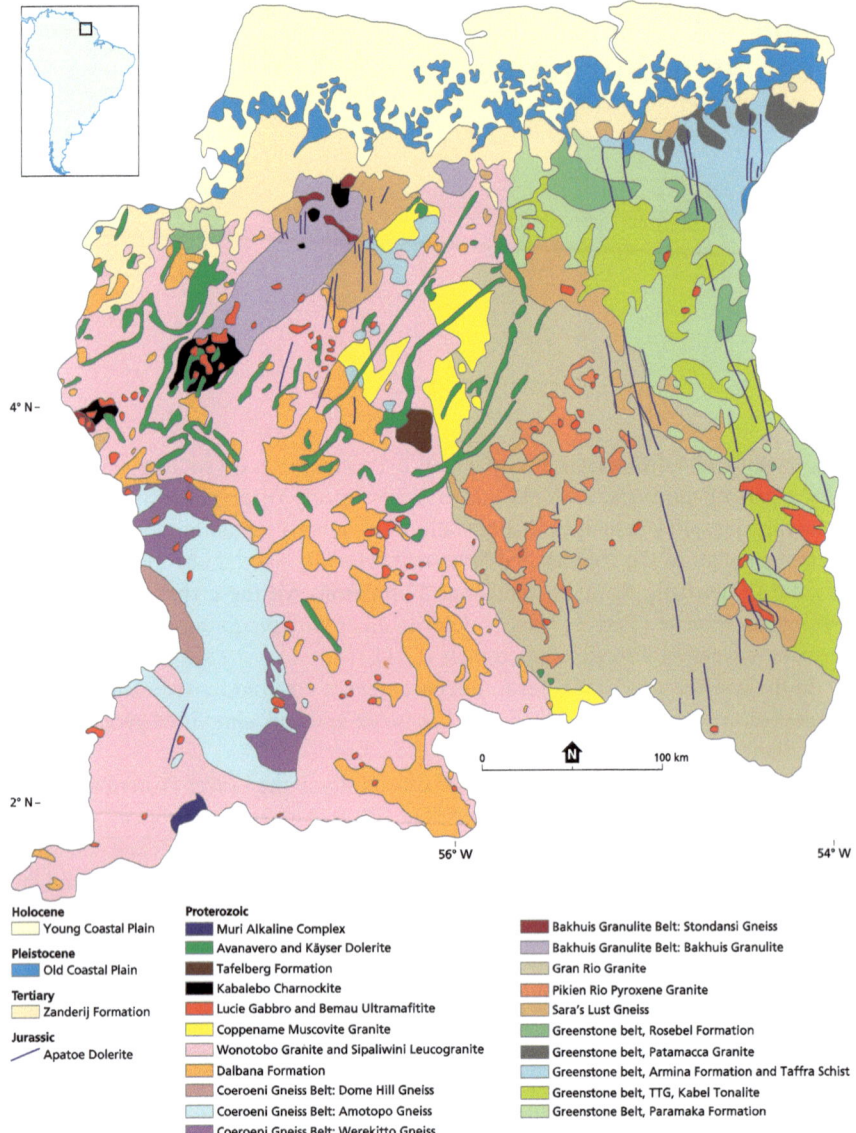

Fig. 2.1 Simplified map with modern data (Kroonenberg et al. 2016, courtesy of NJG–Cambridge University Press)

masses of tuff-lavas, followed by granitoid and gabbroic intrusions in a vast area in central-southern Suriname.

The intrusive stage was followed by a period of cratonisation of the shield, block-faulting, erosion and uplift. Cratonisation was completed before the deposition of the continental Roraima Formation sandstones and pyroclastics, and before the first wave of dolerite intrusions, both occurring in the Middle Proterozoic. The subhorizontal, well-bedded Roraima sandstones and conglomerates occur in the center of the country.

The intrusion of the Permo-Triassic Apatoe dolerites marks the beginning of the separation of South America and Africa in the Mesozoic at 230 Ma. This separation makes a relevant comparison with West Africa's Man-Leo shield and its greenstone belts, including the Birimian Supergroup and their mineral endowment, which is discussed in Chap. 5 on the subject of gold. Gold mineralisation is concentrated in the Trans-Amazonian greenstone belts and in the Mesoproterozoic platform covers. Diamonds originate from Trans-Amazonian volcaniclastic ultramafic rocks (French Guyana) and the Roraima Formation (Guyana), Iron and manganese deposits are formed by lateritic weathering of banded iron formations, gondites and carbonate protores, while bauxite caps are developed on several types of mafic and feldspathic rocks (Kroonenberg et al. 2019).

2.1 Precambrian: Stratigraphy and Formations

The stratigraphic succession observed in Suriname is, from the bottom upwards:

- Volcanic-sedimentary series (Paramaka Formation), which is composed of a basalt sequence of tholeiitic metabasalts, geochemically transitional between Archaean basalts and modern ocean floor basalts associated with metagabbros, and is followed by meta-andesites to metarhyolites; intercalations of metacherts and phyllites increase towards the top.
- Volcaniclastic metagreywacke and phyllite series (Armina Formation).
- Meta-arenite to metaconglomerate series, equally of volcaniclastic origin (Rosebel Formation) (Bosma et al. 1983).

The types and proportions of the volcanic and sedimentary rocks in the greenstone belt (Marowijne Group), and elsewhere on the Guiana Shield, are similar to those of the Canadian Archaean, but differ in these respects from the more mafic-ultramafic belts of the Australian, Indian and southern African Archaean (Gibbs and Barron 1993; cf. Veenstra 1978).

Table 2.1 shows the stratigraphic table and geochronology of the basement of Suriname, simplified after Kroonenberg et al on "Paleoproterozoic evolution of the Guiana shield in Suriname: a revised model" 2016, and based on the most recent geochronological data from samples.

Table 2.1 The stratigraphic table and geochronology of the basement of Suriname, simplified after Kroonenberg et al. (2016)

Time scale	Era	Event	Rock units	Age (Myr)
−100	Cenozoic		Bauxitization	10 and 60
−200	Mesozoic 66–252			
			Apatoe dolerite	200
−300				
−400	Paleozoic 252–541			
−500				
−600				
−700				
−800	Neo proterozoic 1000–541			
−900				
−1000				
−1100			Mylonites	1100–1300
−1200	Meso-proterozoic 1000–1600			
−1300				
−1400				
−1500				
−1600			Kayser and avanavero dolerites	1500–1800
−1700				
−1800			Roraima formation	1800–1900
−1900		Trans-amazonian orogeny	Coeroeni gneiss belt/Coeroeni group	1980–2000
−2000	Paleo-proterozoic 2500–1600		Bakhuis granulite belt/Falawatra group	2050–2080
−2100			Marowijne greenstone belt	2090–2180
−2200				
−2300				
−2400				
−2500				

References

Bosma W, Kroonenberg SB, Maas K, De Roever EWF (1983) Igneous and metamorphic complexes of the Guiana Shield in Suriname. Geologie en Mijnbouw 62:241–254

Bosma W, Kroonenberg SB, Van Lissa RV, de Roever EWF (1984) An explanation of the geology of suriname. Mededelingen Geologisch Mijnbouwkundige Dienst Suriname 27:31–82

Gibbs AK, Barron CN, (1993) The geology of the Guyana Shield–Oxford monographs on geology and geophysics, vol 22. Oxford University Press, New York, pp 246

Goldfarb RJ, Andre-Mayer A-S, Jowitt SM, Mudd GM (2017) West Africa: the world's premier paleoproterozoic gold province. Econ Geol 112:123–143

IJzerman R (1931) Outline of the geology and petrology of Suriname–Thesis Utrecht, Kemink & Zn, Utrecht/Martinus Nijhof, The Hague, 519 pp

Kroonenberg SB, de Roever EWF, Fraga LM, Reis NJ, Faraco T, Lafon J-M, Cordani U, Wong ThE (2016) Paleoproterozoic evolution of the Guiana Shield in Suriname: a revised model. Neth J Geosci 95(4):491–522

Kroonenberg SB, Mason PRD, Kriegsman L, Wong ThE, De Roever, EWF (2019) Geology and mineral deposits of the Guiana Shield. Mededeling Geologisch Mijnbouwkundige Dienst Suriname 29:111–116

Kroonenberg SB (1976) Amphibolite facies and granulite-facies metamorphism in the Coeroeni-Lucie area, southwestern Surinam. Geologisch Mijnbouwkundige Dienst Suriname Mededeling 25:109–289

Texeira JBG, Misi A, da Gloria da Sila M (2007) Supercontinent evolution and the Proterozoic metallogeny of South America. Gondwana Res 11(3):346–361

Veenstra E (1978) Petrology and geochemistry of Sheet Stonbroekoe–Sheet 30, Suriname, Thesis Amsterdam, 161 pp. Also published (1983) in Mededelingen Geologisch Mijnbouwkundige Dienst Suriname 26, 138 pp

Chapter 3
A Brief History of Exploration Work

In 1943, the Geological and Mining Service (GMD) was founded, and exploration efforts were stepped up. Before then, surveying was restricted to traverses of rivers and large creeks on geological field missions. The highlight of the earlier period of exploration was IJzerman's "Outline of the Geology and Petrology of Suriname," published in 1931, which covered all of Suriname's territory.

Subsequently, fieldwork commenced in areas accessible by rail and river from Paramaribo and expanded eastwards and southwards, exploring the coastal plains and the Precambrian basement by following the courses of major rivers. Aerial photographs covering the whole country were taken and analysed.

In 1955, Operation Grasshopper was implemented in order to boost the development of natural resources. Seven airstrips were constructed in strategic locations, in order to allow access to remote regions. An extensive airborne magnetic-radiometric survey was made, covering the Precambrian shield at 150 m altitude and 1,000 m line spacing.

In the period up until 1966, an extensive geophysical survey program was carried out. The United Nations supported this project under the name of Surinam Mineral Survey, from 1961 to 1965. The program included airborne magnetic and electromagnetic surveys, geophysical and geological fieldwork, diamond core drilling and soil sampling. The airborne geophysical surveys revealed several hundreds of partly coinciding magnetic and EM anomalies. Thanks to combined geophysical and lithological properties. anomalies were grouped. including 169 anomalies of first, second and third priority. Of these, 48 anomalies were investigated in the field between 1961 and 1966, by magnetic and electromagnetic surveys (Bosma and Lokhorst 1975).

As a follow-up to the airborne geophysical investigations, the GMD started in 1974 with a base metal exploration project on which geophysical, geological and geochemical methods were integrated with an emphasis on geophysics. The De Goeje type gabbroic bodies were of special interest, in view of indications for associated sulphide mineralisations. The main methods used were combined induced polarization/resistivity and magnetics. Anomalies in 8 areas, all except one on the Bakhuis Horst (the Maratakka anomaly is located on the savannah belt north of

M. Keersemaker, *Suriname Revisited: Economic Potential of its Mineral Resources*, SpringerBriefs in Earth Sciences, https://doi.org/10.1007/978-3-030-40268-6_3

Bakhuis), were investigated, including the use of drilling. In three areas follow up was recommended including one area with copper/phosphate mineralization, which was the only area where further investigations were carried out.

3.1 Geochemical Survey

In 1961, GMD started with geochemical exploration, which initially consisted of detailed soil sampling over mineral indications and geophysical anomalies, and which included taking stream sediment samples incidentally. Gradually, this became more important, also thanks to Doeve (1966), who recommended geochemical exploration in order to locate metallogenic anomalies. During the period 1964–1972, about 11,000 geochemical soil samples were investigated, mainly from areas showing an aero-geophysical anomaly, either magnetic or electromagnetic. In some cases, sampling was carried out after the discovery of ore minerals, in order to establish the possible extension of the deposit. All metal determinations were carried out in the laboratory of the GMD (Oosterbaan 1973).

In 1972, a regional geochemical exploration program commenced, with systematic collection of stream sediment samples. Teams of geologists combined geological mapping with collecting stream sediment samples, using a low-density pattern.

From 1972 to 1974, a regional test survey was carried out over an area of around 2,900 km^2 divided into blocks of 3 km by 3 km. In each of these 9 km^2 squares, a stream sediment sample was collected at the most representative drainage site. A soil sample was also taken at 100 m distance uphill from the stream sediment location. During subsequent regional stream sediment surveys, this sampling pattern was abandoned as follow-up from anomalies was difficult, due to uncertainty regarding the exact location of the catchment areas of the samples. Simultaneously, soil sampling was also stopped, as it provided no significant additional information (Dahlberg 1982).

Hundreds of thousands of samples have been analysed by X-ray fluorescence spectrometry, atomic absorption spectrometry, emission spectography and wet chemical methods. Samples were routinely checked for Cu, Ni, Co, Cr, Pb, Zn, Mn, Mo, Au and Ag (De Vletter 1984).

The geochemical analysis resulted in the location of various mineral occurrences or anomalies. Figure 3.1 shows the areas covered by geochemical surveys.

Several maps have been produced with indications of mineral occurrences resulting from above-mentioned surveys. The works of Doeve in 1966 and Bosma, Ho Len Fat en Welter in 1972 were followed by the metallogenic map reproduced below, compiled in 1976 by Dahlberg. De Vletter published a simplified version in 1984, which is shown in Fig. 3.2 below.

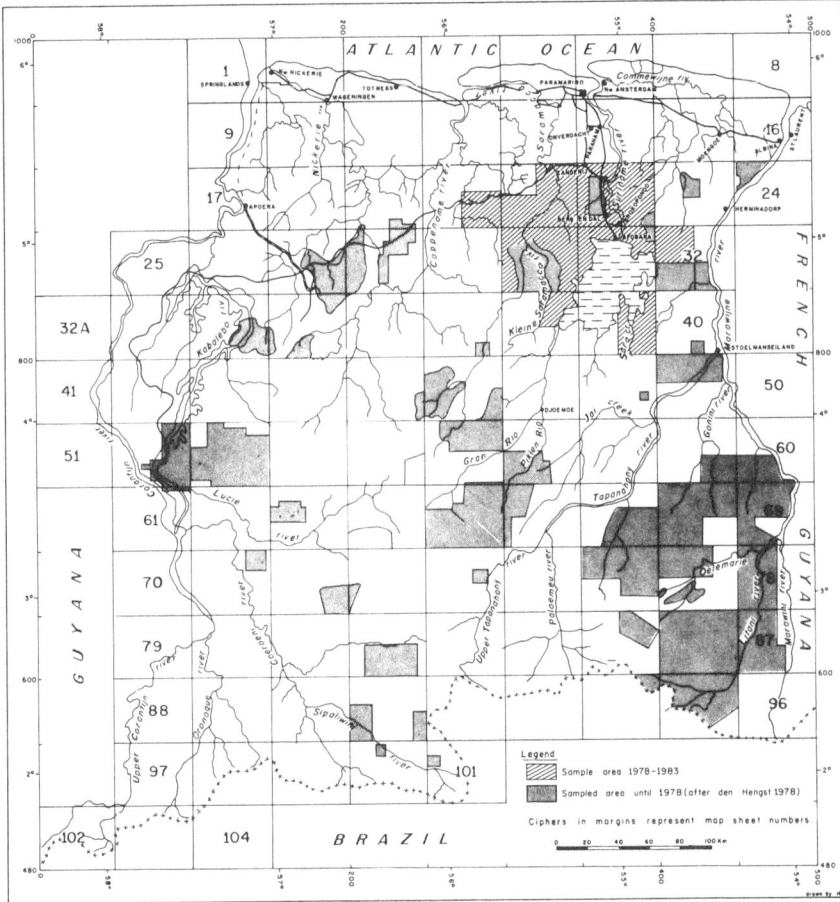

Fig. 3.1 Areas subject to geochemical surveys (De Vletter 1984)

3.2 Exploring for Bauxite and Gold

Historically, bauxite and gold have received extra attention, as mining companies (including small-scale miners) invested in exploration in order to develop mining projects, and later to continue and expand their operations.

Billiton Maatschappij Suriname (a subsidiary of Billiton and BHP) and Suralco (a subsidiary of Alcoa) were active in Suriname's bauxite sector for 70 and 100 years, respectively, until their recent closure of active operations in 2009 and 2015. In order to sustain their operations, all known bauxite deposits have been explored extensively, within their permit areas between Saramacca River and Moengo Area. A large number of drill holes (mainly by auger drills) have been sampled, including sites on Lely Mountains, on Nassau Mountains, and on Bakhuis Mountains. Those

Economic geology and mineral potential of Suriname.   S U R I N A M E : Mineral   occurrences

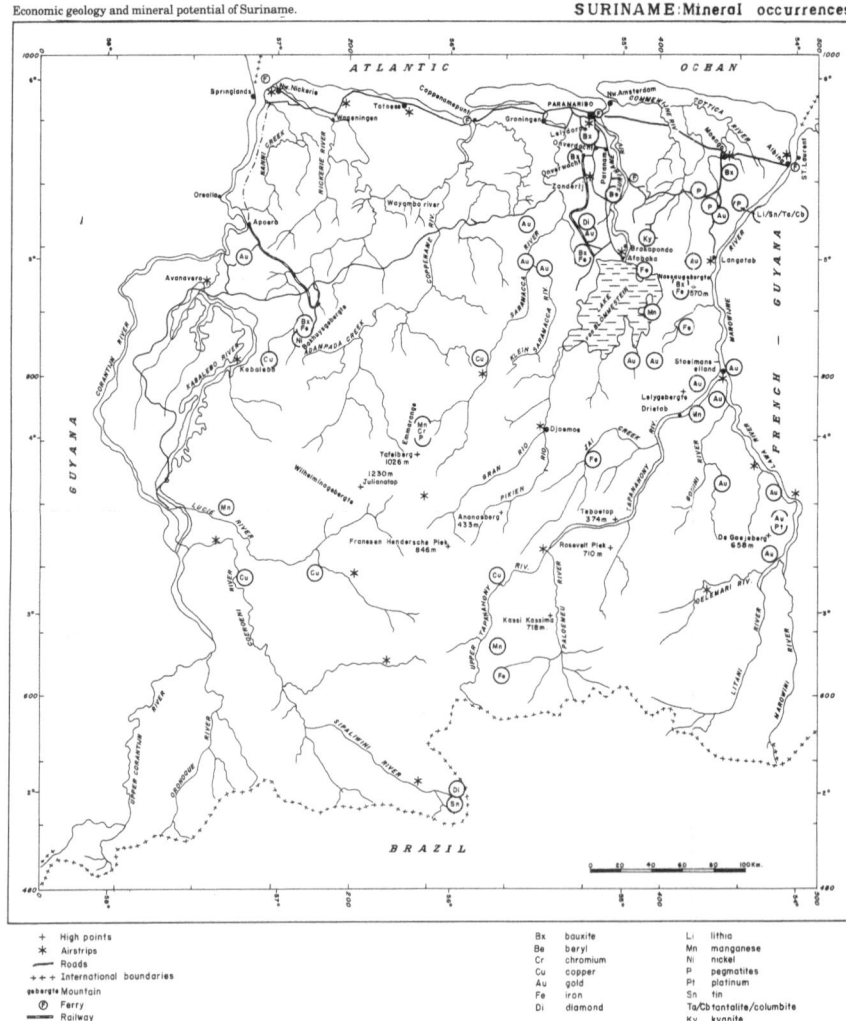

Fig. 3.2   Mineral occurrences map of Suriname (De Vletter 1984)

regions are not (as yet) developed, the implications of which will be further discussed in Chap. 6.

Between 1952 and 1974, a number of concessions on the coastal plain were explored by the Guiana Exploration Company, Reynolds Metals Company and the GMD, as described by Van Lissa (1975). More than 1,000 holes were drilled in these areas without success, despite the presence of about 100 separate bauxite occurrences in a bauxite belt of 320 km long and 20–30 km wide, in the coastal plain of Guyana across the Korantijn River. Van Lissa suggested that this apparent discrepancy may have occurred because some areas may not have been drilled deep enough, as will be discussed in Chap. 6.

As for gold, Iamgold (initially Cambior) and Newmont explored and developed the Rosebel and Merian mines (see Fig. 5.1), which started in 2004 and 2016, respectively. Both companies have carried out extensive diamond drilling campaigns and aerial surveys in their areas of interest, to meet resource and reserve reporting standards and to explore future growth options.

References

Bosma W, Lokhorst A (1975) Geophysical, geological and geochemical characteristics of some De Goeje typy gabbroic bodies. Mededelingen Geologisch Mijnbouwkundige Dienst Suriname 23:176–193

Dahlberg EH (1976) The metallogenic map of Suriname, explanatory note. Mededelingen Geologisch Mijnbouwundige Dienst Suriname 24:10 pp

Dahlberg EH (1982) Geochemical exploration within the Precambrian terrain of Suriname (Guiana Shield). In: The development potential of Precambrian mineral deposits–UNDP, Natural Resources and Energy Division. Pergamon Press, Oxford, 397–407

De Vletter DR (1984) Economic geology and mineral potential of Suriname. Mededelingen Geologisch Mijnbouwkundige Dienst Suriname 27:11–30

Doeve G (1966) Delfstoffen in Suriname. Mededeling 15 van Geologisch Mijnbouwkundige Dienst van Suriname, 94–107

Oosterbaan WE (1973) Geochemical exploration in Suriname–a review. Mededelingen Geologisch Mijnbouwkundige Dienst Suriname 22:1103–118

Van Lissa RV (1975) Review of bauxite exploration in the coastal plain of Suriname. Mededeling Geologisch Mijnbouwkundige Dienst 23:250–254

Chapter 4
Modeling Economic Potential: Minimal Resource Parameters

As described by Lane, in his book "Economic Definition of Ore", minerals in the ground have no explicit value. It is not until they have been mined, processed and delivered to a customer that any value is realised. A mineralised body should therefore be regarded as a possible opportunity for development (Lane 1991).

Two methods are considered to provide an indication of the economic potential for the various commodities that have been identified in the course of exploration efforts described in Chap. 3.

The first method is comparative: it looks at size and grade of deposits known across the world which are actually being mined currently, or in the not too distant past, or which are in an advanced feasibility study phase and therefore form a good reference.

The second method is projective: it is based on a formula that calculates the minimum size and grade of a resource that would be required to justify an estimated amount of investment capital. In this approach, input of rough capital estimates are used, based on conceptual models for mining, processing and transportation. The formula will be referred to as the Proven & Probable-formula (or P&P-formula), which is explained further below.

4.1 Concept of the Proven and Probable-Formula

To evaluate the potential of deposits for further development into a mine with economic returns, a Proven and Probable reserves formula was developed. This formula calculates the minimum size and grade of a resource required to justify an estimated amount of investment capital.

For gold, the formula is calculated for three capacities: 3 Mtpa, 6 Mtpa and 12 Mtpa plant throughput, the latter being the current size of Rosebel and Merian's annual plant throughput. These are Suriname's only large-scale gold producing mines. For each case the formula has two major components: a capital estimate and a cashflow profile, both of which are discussed below.

© The Author(s), under exclusive license to Springer Nature Switzerland AG 2020
M. Keersemaker, *Suriname Revisited: Economic Potential of its Mineral Resources*,
SpringerBriefs in Earth Sciences, https://doi.org/10.1007/978-3-030-40268-6_4

4.2 Investment Capital Estimates

The costs required in order to access an area (deep) inside the Marowijne Greenstone belt area for mine development are significant, and depend in large part on the distance between mine and existing road infrastructure. The Marowijne greenstone belt area can roughly be divided into three zones, of which the north part has been divided by Iamgold and Newmont as areas of strategic interest. For this publication, the focus is on the remaining part in the south, as divided into:

- Marowijne Mid-Zone between latitudes 5° and 4° including Lely Mountains at about 100 km distance from existing road infrastructure.
- Marowijne South-Zone between latitudes 4° and 3° including Benzdorp at about 200 km distance.

Depending on the location of the deposit, an investment in a heavy-duty road of ~100 or ~200 km is required. At this conceptual stage, such a road through dense forest with sufficient capacity for safe industrial transport is estimated to cost 1 M US$ per kilometer.

4.3 Mine Development

The following breakdown of mine development cost items is used:

- Processing plant (crusher, mill, gravity circuit)
- Mine Infrastructure (workshops/warehouse/offices/camp/water/sewage)
- Power plant (Heavy Fuel Oil-based generator units)
- Earthworks (Pre-stripping, mine roads, tailings dam)
- Access road (to local network: 100 and 200 km respectively)
- Railway line (not required for gold)
- Port facility (not required for gold)
- Mobile equipment and spares
- Exploration and study costs
- Indirect project costs as a percentage of above, for Engineering, Procurement, Construction Management (10%) and distributables (10%)
- Project owners' cost (20%)
- Contingency (15%).

4.4 Cashflow Profile

The following items are used as parameters in the formula:

a. Commodity price
b. AISC (All-in sustaining cost) excluding royalties (Whelan 2013)
c. Royalties & Corporate tax

d. Profit and Risk margin (at 20% of the gold price)
e. Capital per unit of sale
f. Saleable production
g. Proven and Probable reserve.

Starting with gold, in the following chapter, parameter (a.) is the price of gold per ounce, components b-d are deducted. The balance of $a - (b + c + d) = e =$ the amount in US dollar available per ounce for investment in mine, access and plant. The Proven and Probable reserve required to support such investment (f) is calculated by dividing the total capital estimate by the capital available per ounce, to give the number of ounces produced for sale. The value of (f) is the actual saleable production recovered from the reserve (g).

References

Lane KF (1991) Economic definition of ore. Mining Journal Books, 149 pp
Whelan T (2013) All-in sustaining costs and all-in costs. Ernst & young presentation at Americas mining and metals forum

Chapter 5
Gold

The Guiana Shield is among the least known among Precambrian terrains, because access is relatively difficult, and furthermore, intense weathering has limited bedrock exposure. This situation has significantly improved in recent times, when shallow in situ gold occurrences attracted exploration and mining companies to commence geological programs aimed at better understanding the geology and the mineral deposits of the Shield (Voicu et al. 2001).

In terms of modern-day exploration, the Marowijne greenstone belt situated in Eastern Suriname has been under-explored, in comparison to neighbouring countries and also compared to the greenstone belts on the Man-Leo shield in West-Africa. Programs carried out by Iamgold and Newmont were limited to the northern part of the belt thanks to available road access. A long history of activities by small-scale miners in more remote areas indicate occurrences of (alluvial) gold in various locations distributed over the greenstone belt area as indicated in Fig. 5.1. The easily recoverable alluvial and eluvial gold has probably already been discovered (De Vletter 1984), but further prospecting targets are still available.

Based on proven potential in the north, which has led to the development of Rosebel and Merian mines, as well as on the mineral endowment of and actual mineral output from the associated Man-Leo Shield, it would be expected that Suriname's Marowijne greenstone belt hosts more mineral deposits with economic potential further south in less accessible areas.

This chapter provides an answer to the question: what must the minimum grade and size of a gold deposit in the Marowijne belt be, to make exploitation on a larger scale potentially feasible. The answer to this question can provide the basis for decisions to invest in exploration and further feasibility studies. The starting point lies at the economic end of a mineral resource value chain and reverses back to the geology.

© The Author(s), under exclusive license to Springer Nature Switzerland AG 2020
M. Keersemaker, *Suriname Revisited: Economic Potential of its Mineral Resources*,
SpringerBriefs in Earth Sciences, https://doi.org/10.1007/978-3-030-40268-6_5

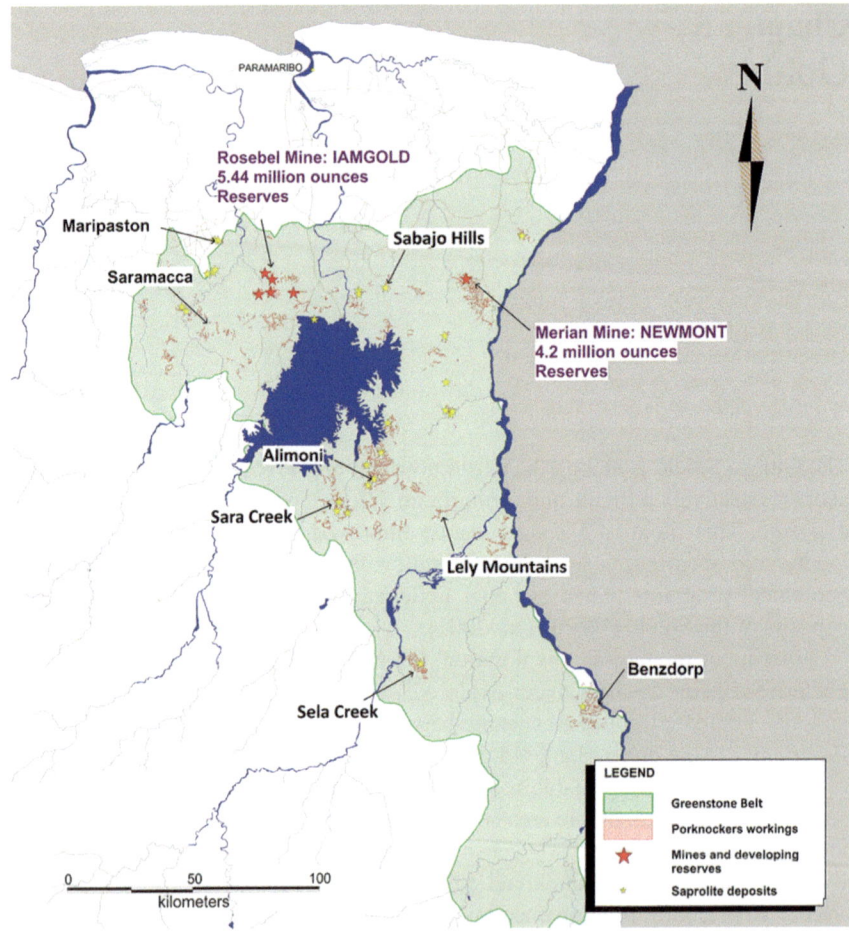

Fig. 5.1 Marowijne Greenstone belt (Kioe-A-Sen et al. 2016 courtesy of NJG—Cambridge University Press)

5.1 West Africa as Reference

The Guyana shield part of the Amazon craton and the West Africa craton have a similar history of crustal growth and collision events that define the Transamazonial orogeny in South America and the Eburnean orogeny in West Africa. They also contain at one time contiguous Paleoproterozoic greenstone belts stretching from eastern Venezuela and northeastern Brazil to Mali, Burina Faso and Nigeria as shown in Fig. 5.2, and they share the same metallogenic features with orogenic gold formation concentrated at 2.1 Ga (Goldfarb et al. 2017).

The Birimian sequences of the West African Man-Leo shield in particular are highly endowed in minerals specifically gold but also include iron ore, bauxite,

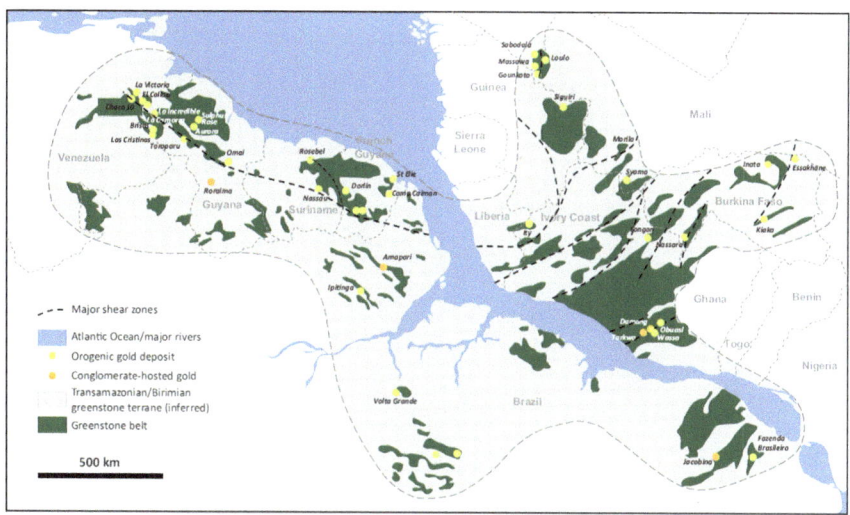

Fig. 5.2 Reconstruction of the Guyana and West African shields with locations of greenstone belts and gold deposits (Goldfarb et al. 2017)

diamonds, manganese, phophate and base metals. These Birimian belts cover parts of Mali, Ghana, Cote d'Ivoire, Burkina Faso and stretch into Senegal and Mauretania in the West and North over an area of about 350,000 km^2 (Nyame 2013; Markowitz et al. 2015).

For at least 1,500 years, gold has determined West Africa's history and economy, and in the period between 400 and 1500, the region was the world's leading gold exporter.

The Obuasi deposit in Ashanti region of Ghana is the largest gold deposit, with >2,500 tonnes of gold. Mining started here in the 1890s and continued until operations were suspended in 2014. Following development of access to deeper levels, production is expected to re-start by the end of 2019.

A compilation carried out by Goldfarb et al. in 2017 indicated almost 10,000 tonnes of contained, compliant gold resources hosted by 138 West African gold deposits, almost all classified as orogenic and with an average grade of 1.68 g/t.

In 2013, Natural Resources Holdings published the Global Gold Mine and Deposit rankings of all 580 mines and deposits >1 Moz contained gold. Of those 580, 50 mines and deposits are located on the Man-Leo Shield with a total of 243 Moz contained gold. In comparison, the Guiana shield hosts 10 mines and deposits with a combined total of 85 Moz. The difference is significant and suggests there is hidden potential certainly for gold, but also for other commodities. Gold is the main revenue-earner of the Birimian, but the same area produces copper from copper/gold porphyry style deposits, as in Burkina Faso, lead-zinc from volcano-genic massive sulphide deposits, iron ore from skarns and Banded Iron Formation, manganese, bauxite from young enrichments and diamonds, as shown in Fig. 5.3.

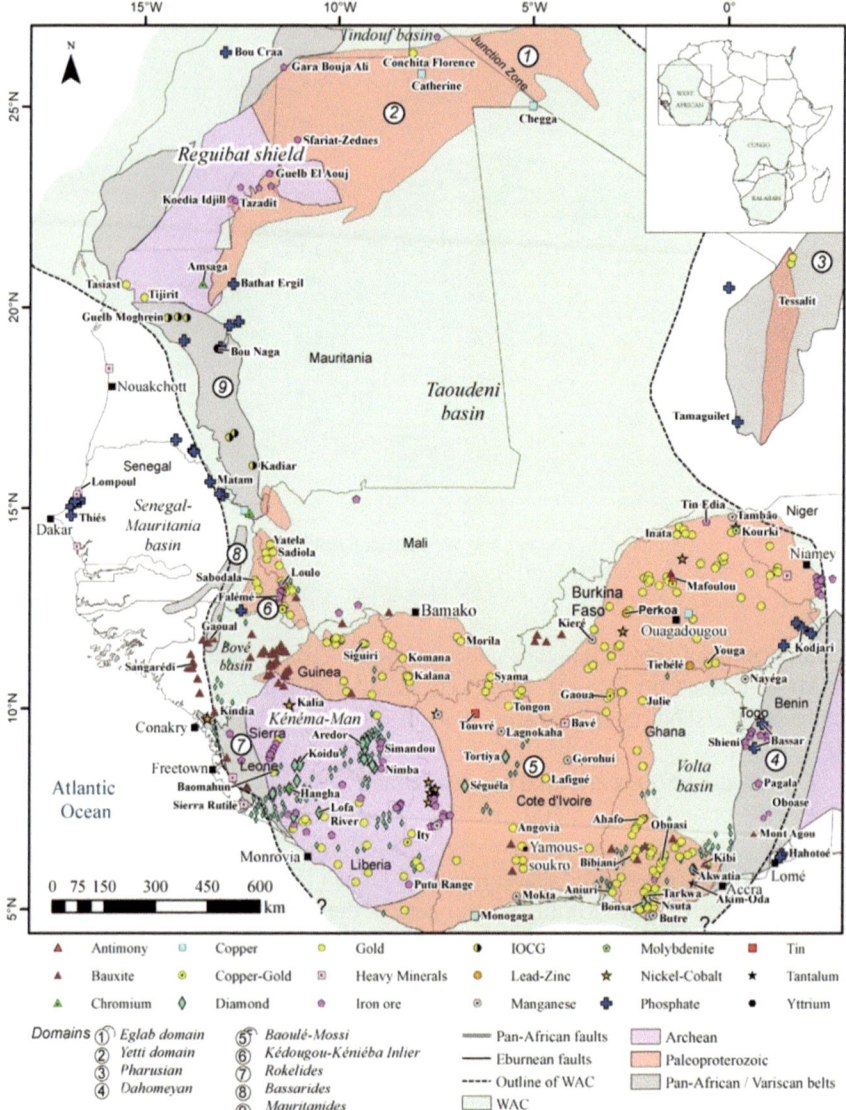

Fig. 5.3 Mineral endowment of West Africa (Markwitz et al. 2016)

5.2   Gold-Producing Operations in Suriname

In the northern part of the Greenstone belt, Iamgold Corporation and Newmont Mining Corporation have developed two world-class gold deposits since the start of this century: Rosebel and Merian. In the Global 2013 Gold Mines and Deposits rankings

(Natural Resources Holdings 2013) they were ranked #105 and #182 respectively, in contained ounces at the time.

5.2.1 Rosebel Gold Mines

The Gros Rosebel area, located about 80 km south of Paramaribo has produced a large part of the country's recorded gold production since the last quarter of the 19th Century, when gold was first discovered in the area (Wong et al. 1998). A Canadian company, Golden Star, reached a mineral agreement in 1994 in joint venture with Cambior, but it was not before 2003 that Rosebel Gold Mines was developed by Cambior alone. The project had the following characteristics (Cambior Annual Report 2003):

Probable reserves: 47 million tonnes at 1.6 g/t Au equivalent of 2.4 million ounces contained gold:

- Planned output 220,000 oz per annum at 215 $/oz
- Assumed gold price: 400 $/oz
- Projected mine life: 10 years
- Initial capital cost: 95 million US$
- Mining rate: 16.2 Mt/year
- Milling rate: 4.6 Mt/year.

Rosebel Gold Mines (RGM) started production in 2004, and has produced about 5 Moz since then with an annual production of about 300,000 oz in 2018, at a total cash cost of 831 $/oz and an all-in sustaining cost of 1,006 $/oz. Iamgold acquired RGM and Cambior in 2006.

In their latest 2018 annual report, Iamgold presented proven reserves of 30 Mt at 0.6 g/t and probable reserves of 101 Mt at 1.0 g/t, which equals 3.8 million ounces of contained gold. The mine infrastructure and milling capacity has been expanded over time to 16 Mtpa, at an estimated total cost of about 945 million US$ at today's value (Iamgold Annual reports 2006–2018).

5.2.2 Merian

Newmont's Merian mine started production in the fourth quarter of 2016. The estimated annual production is in the range of 450–500,000 oz per year.

Newmont published reserves of 134 Mt at 1.2 g/t, which equals 4.8 million ounces contained gold (Newmont investor presentation 2015). Assumed gold price is 1,300 $/oz.

The project was completed in 2016 at a capital investment of 869 M US$ (Newmont Annual Report 2016).

Fig. 5.4 Newmont Mining's Merian 2 pit

Expected all-in sustaining cost per ounce ranges from 650 to 750 US$. Actual AISC for 2017 was 572 US$/oz sold as per Newmont's 2017 annual report:

- Expected mining rate 50 Mt/year
- Expected milling rate 12 Mt/year.

Figure 5.4 shows a view of one of Merian's two primary pits with gold-bearing weathered saprolite exposed. The Merian 2 pit shown below will ultimately be 2.5 km long and 500–700 m wide with a final depth of 260 m. The saprolite layer is about 100 m thick overlaying fresh rock and contains a series of sheeted quartz veins with a NE dip of 40–50° (Kioe-A-Sen et al. 2016). Most veins have a thickness ranging between 10 and 25 m resulting in a strip ratio of ~3.

5.3   Economic Model for Gold

For the input values in the model Industry sources (BHP Billiton, Iamgold, Newmont and others) were used in addition to values published from the Hounde project in Burkina Faso below, which is used as reference case.

5.3.1   Hounde Gold Project—Burkina Faso

Endeavour started construction in 2016 of its Hounde mine in Burkina Faso, an up-to-date low-cost gold mining project situated on the Birimian belt. This project forms a good example and benchmark for its comparative grade, size, cost, remoteness, process (carbon-in-leach) and deposit style. The project delivered a 3 Mtpa mine and plant with a 10 year mine life.

The company published a breakdown (at www.endeavourmining.com) of their approved capital investment in the project. In October 2017, Endeavour announced that the construction was completed at 15 million US$ within budget, thus only half of the contingency amount was used.

Table 5.1 Capital estimates. The numbers in red in the bottom row are the total capital estimates in million US$ used as input in the P&P-formula

Capital cost structure	3 Mtpa	3 Mtpa		6 Mtpa		12 Mtpa	
	Hounde	Mid	South	Mid	South	Mid	South
Processing plant	57	60	60	110	110	160	160
Maintenance/Warehouse/Camp	30	30	30	40	40	50	50
Power plant	17	25	25	45	45	65	65
Pre-stripping and tailings dam	28	30	30	50	50	80	80
Access roads		100	200	100	200	100	200
Equipment and spares	47	50	50	100	100	160	160
Exploration		20	20	30	30	40	40
Subtotal	179	315	415	475	575	655	755
Indirects:							
EPCM@10%	17	32	42	48	58	66	76
Distributables @ 10%	21	32	42	48	58	66	76
Owners' cost @ 20%	83	63	83	95	115	131	151
Contingency @ 15%	28	47	62	71	86	98	113
Total in M USD	328	488	643	736	891	1,015	1,170
Rounded input for P&P model		490	650	740	900	1,020	1,170

The Hounde project uses high voltage power supply which will not apply to the Suriname cases. For Marowijne the estimates are based on installed power of 20, 40 and 65 MW for 3, 6 and 12 Million tonnes per annum throughput respectively at an estimated cost of 1 M US$ per MW (Industry source).

The resulting capital estimates for the different options are listed in Table 5.1.

5.3.2 Gold Price

As shown in Fig. 5.5 the gold price has been fairly stable over the last 5 years. The formula assumes an average of 1,300 US$/oz. This price is comparable to what gold companies assume in their forecasts (Industry sources: Annual reports Iamgold, Newmont, Endeavour Mining). There are many opinions on what the future holds for the price of gold, but due to the cost of operations (see below) coupled with an increase in global uncertainties, the price is not likely to be much lower in the foreseeable future.

Fig. 5.5 Ten-year gold price (with January datapoints from Infomine.com)

5.3.3   Operating Costs and All-in Sustaining Costs

The World Gold Council (WGC) defines all-in sustaining cost or AISC (see guidance notes in annual reports and Ernst & Young brief 2013). This non-GAAP (General Association of Accountants Practice) definition is often used in published statements of gold companies. It claims to reflect the true cost of operating gold mines incurred in the complete mining life-cycle from exploration to closure.

Note that non-sustaining costs are costs incurred in relation to new operations and costs related to major projects at existing operations, where these projects will materially increase production. This parameter AISC is used here with the assumption that it forms a true reflection of the operating cost for the given model.

The AISC includes the "old" cash costs:

- Direct mining costs
- Royalties
- By-product credits
- Gains realised under currency hedge
- Other.

In addition, AISC includes:

- General administrative costs
- Mine development expenditures
- Sustaining capex
- Mine on site exploration
- Rehabilitation/accretion and amortisation.

Newmont and Iamgold publish their anticipated and past AISC which is used as input in the formulae. Credits from other commodities are not considered for this Marowijne analysis.

5.3.4 Royalties and Taxes

The existing gold mines in Suriname are publishing (historic and anticipated) royalty and tax payments (Annual reports Cambior, Iamgold and Newmont). Both Rosebel and Merian have custom-made agreements with the Government of Suriname for royalties, and the corporate tax rate is 36%. It is assumed that a future mine in Marowijne would be liable to pay similar amounts per ounce.

Based on historic values for RGM and projected values for Merian (source: Annual reports Iamgold, Newmont and Government of Suriname), the royalties and corporate income tax average 11.5% of the gold price in total, and are evenly distributed between the two.

5.3.5 Proven and Probable Reserves

The total estimated investment capital required for the 6 options under review divided by the available capital per ounce gives the total number of ounces needed. These are the number of total ounces for each option that a deposit would have to contain, in order to be able to afford payback of the calculated investment. It is assumed that actual saleable production is 94% of reserves, which is the reported recovery by Iamgold as per 2018.

5.3.6 Results

The Table 5.2 shows the results of the P&P-formula application with the above input: the left side for the South-zone and the right side for the Mid-zone of the Marowijne greenstone belt. Three plant capacity scenarios were calculated for both geographic zones.

Operating Costs (AISC) are the same for all scenarios. It is assumed that benefits from economy of scale are reduced by a relative increase in general and administrative cost for each of the large capacity options. The formula is intended to provide a good indication of whether further studies are justified. More detailed engineering studies and exploration drilling will then provide the basis for an informed investment decision.

A 10 year mine life was assumed as a reasonable horizon for investment perspective and payback time. Longer life scenarios, where discount factors have a major impact on long-term cash flows, have not been considered for this research.

The results show that there are several options ranging from 5 Moz comparable to Merian and Rosebel in the Mid Zone to a smaller scale, higher grade deposit of 3.2 Moz in the South Zone. These are comparable to Randgold's Tongon deposit in Cote d'Ivoire and Resolute's Syama deposit in Mali, both situated on the Birimian

Table 5.2 Overview of P&P-formula calculations and results

P+P Formula	STH	3 Mtpa
	Capital	650 million US$
	gold price	1300 US$/oz
Less	AISC ex-royalties	675
		625
Less	Royalty&tax	150 11.5% of gold price
		476
Less	Profit&risk	260 20% of gold price
Balance:	Capital/oz	216 US$
Saleable production in 10 yr		3.0 Moz
Minimum P+P reserve		3.2 Moz

P+P Formula	MID	3 Mtpa
	Capital	490 million US$
	gold price	1300 US$/oz
Less	AISC ex-royalties	675
		625
Less	Royalty&tax	150 11.5% of gold price
		476
Less	Profit&risk	260 20% of gold price
Balance:	Capital/oz	216 US$
Saleable production in 10 yr		2.3 Moz
Minimum P+P reserve		2.4 Moz

P+P Formula	STH	6 Mtpa
	Capital	900 million US$
	gold price	1300 US$/oz
Less	AISC ex-royalties	675
		625
Less	Royalty&tax	150 11.5% of gold price
		476
Less	Profit&risk	260 20% of gold price
Balance:	Capital/oz	216 US$
Saleable production in 10 yr		4.2 Moz
Minimum P+P reserve		4.4 Moz

P+P Formula	MID	6 Mtpa
	Capital	740 million US$
	gold price	1300 US$/oz
Less	AISC ex-royalties	675
		625
Less	Royalty&tax	150 11.5% of gold price
		476
Less	Profit&risk	260 20% of gold price
Balance:	Capital/oz	216 US$
Saleable production in 10 yr		3.4 Moz
Minimum P+P reserve		3.6 Moz

P+P Formula	STH	12 Mtpa
	Capital	1170 million US$
	gold price	1300 US$/oz
Less	AISC ex-royalties	675
		625
Less	Royalty&tax	150 11.5% of gold price
		476
Less	Profit&risk	260 20% of gold price
Balance:	Capital/oz	216 US$
Saleable production in 10 yr		5.4 Moz
Minimum P+P reserve		5.8 Moz

P+P Formula	MID	12 Mtpa
	Capital	1020 million US$
	gold price	1300 US$/oz
Less	AISC ex-royalties	675
		625
Less	Royalty&tax	150 11.5% of gold price
		476
Less	Profit&risk	260 20% of gold price
Balance:	Capital/oz	216 US$
Saleable production in 10 yr		4.7 Moz
Minimum P+P reserve		5.0 Moz

greenstone belt. The Mid-zone of the belt requires slightly lower grades and total contained ounces for economic development. All 6 options would sit in the range of spots 340–150 in the top 580 (Global 2013 Gold mine and deposit rankings). That same range includes 21 deposits on the Man Leo Shield versus 3 on the Guiana Shield: Paul Isnard (French-Guyana), Increible (Venezuela) and Merian. Rosebel should be the number four, following an upgrading of reserves with the inclusion of Saramacca.

These numbers suggest that there is potential for one or more gold deposits in Eastern Suriname that fit this profile. Further prospecting targets include deeper alluvials, areas with fine-grained gold, gold in sulfides and regular vein-type deposits.

References

Annual reports Cambior Inc. (2003–2005)

Annual reports Iamgold Corporation (2005–2018)

Annual reports Newmont Mining Corporation (2015, 2017)

De Vletter DR (1984) Economic geology and mineral potential of suriname. Mededelingen Geologisch Mijnbouwkundige Dienst Suriname 27:11–30

Endeavour Mining Press Release, April 11, 2016. www.endeavourmining.com

Goldfarb RJ, André-Mayer A-S, Jowitt SM, Mudd GM (2017) West Africa: The World's Premier Paleoproterozoic Gold Province, Economic Geology, 112:123–143

Kioe-A-Sen NME, Van Bergen MJ, Wong ThE, Kroonenberg SB (2016) Gold deposits of Suriname: geological context, production and economic significance. Neth J Geosci 95(4):429–445

Markowitz V, Hein KAA, Miller J (2015) Compilation of West African mineral deposits: spatial distribution and mineral endowment. Precambrian Res 274(2016):61–81

Markwitz V, Hein KAA, Jessell MW, Miller J (2016) Metallogenic portfolio of the West African craton. Ore Geol Rev 78:558–563

Natural Resources Holdings: Global 2013 Gold Mine and Deposit rankings—visualcapitalist.com

Nyame FK (2013) Origins of Birimian mafic magmatism and the Paleoproterozoic "greenstone belt" metallogeny: a review. Island Arc 22(4):538–548

Voicu G, Bardoux M, Stevenson R (2001) Lithostratigraphy, geochronology and gold metallogeny in the northern Guiana Shield, South America: a review. Ore Geol Rev 18(3–4):211–236

Whelan T (2013) All-in sustaining costs and all-in costs. Ernst & Young presentation at Americas Mining and Metals Forum

Wong ThE, de Vletter DR, Krook L, Zonneveld JIS, van Loon AJ (eds) (1998) The history of earth sciences in Suriname, 479 pp

Chapter 6
Bauxite

Bauxite is the world's main source of aluminium, and Suriname has a long history of mining and processing this ore for export.

More than 85% of the bauxite mined globally is converted to alumina for the production of aluminium metal. An additional 10% goes to non-metal uses in various forms of specialty alumina, while the remainder is used for non-metallurgical bauxite applications.

Non-metallurgical bauxite is used in:

- cement products
- chemical products
- calcined products.

Calcined products include refractory and abrasive applications, as well as special uses, such as proppants for the oil industry (Hill and Sehnke 2006).

In most commercial operations, as in Suriname's refinery at Paranam, alumina is refined from bauxite by a wet chemical caustic leach process, known as the Bayer process. Alumina is then smelted, using energy-intensive electrolytic reduction, to produce aluminium, which process took place in Suriname in the previous century, before alumina became the sole export earner.

Bauxite grades are generally reported as a range of oxides including Al_2O_3, SiO_2 and Fe_2O_3. Principal mineral components of bauxite include the three main minerals:

- gibbsite (referred to as tri-hydrate)
- boehmite (referred to as mono-hydrate)
- diaspore (same composition as boehmite, but requires higher refining temperature).

Gibbsite is the principal bauxite mineral in all the Suriname deposits, though all deposits contain at least some boehmite (Van Kersen 1956).

The other minerals are impurities, and include:

- clay
- quartz (SiO_2)
- iron oxides (Fe_2O_3)
- titanium oxides (TiO_2).

M. Keersemaker, *Suriname Revisited: Economic Potential of its Mineral Resources*, SpringerBriefs in Earth Sciences, https://doi.org/10.1007/978-3-030-40268-6_6

The major oxides reported are usually referred to as Total Al_2O_3, Total SiO_2, etc., where Total Al_2O_3 = gibbsite, boehmite, diaspore + clay minerals and Total SiO_2—clay minerals (kaolinite), quartz.

Not all of the Al_2O_3 (alumina) is available for making aluminium, as some resides in the clays and is lost during the Bayer process. The total amount of alumina that is extractable in solution from bauxite in the Bayer process at 143 °C is called the Available Alumina (AA or Al143).

Silica is the most commercially important impurity in bauxite. Reactive silica is silica which reacts with comparable amounts of alumina during the Bayer process digestion of bauxite to form insoluble sodium aluminum silicate, and which is lost as a refinery plant residue. Generally, the more silica in the bauxite, the higher the amount of caustic soda is consumed in the refining process, and the higher the loss of alumina to the red mud tailings.

6.1 Geology

The first publications on Suriname's bauxite date back to 1903, and several studies on the bauxite deposits and their origin followed.

The lateritic bauxite deposits of Suriname developed on Cenozoic sediments, in the coastal lowlands and metamorphic rocks of the Precambrian Guiana Shield (Monsels 2016). After the Tertiary uplift of the basal complex in the interior, renewed erosion took place. The erosion products were transported to the coastal areas by rivers, depositing arkosic and subarkosic sands and clays. After peneplanation, which formed in the early Tertiary surface, bauxitisation took place in the more basic rocks in the hinterland and the arkosic-subarkosic sediments of the coastal plain, thus forming plateau bauxite and lowland bauxite, respectively. Large parts of the surface disappeared, during later uplift and erosion, but those covered with bauxite or laterite remained relatively untouched, thus forming the flat-topped Nassau Mountains, Lely Mountains, Brownsberg and Bakhuis Mountains in the interior, and Moengo Hills and Onverdacht in the eastern coastal plains (Bosma et al. 1973). The bauxite in the coastal plains is covered with younger sediments in Onverdacht or surrounded by them, in the case of the Moengo Hills. The lowland bauxite is generally of good quality.

6.2 History

In 1914, an English mining engineer claimed a bauxite discovery in Saramacca district, and acquired gold concession rights, because the mining law did not include the category of bauxite. This could be considered to be the first bauxite concession in Suriname (Wong et al. 1998).

In 1916, Alcoa geologists discovered bauxite in the Moengo Hills area. A mine was developed, and export of bauxite from Suriname started in 1922. In the buildup towards (and during) the Second World War, demand for bauxite from Suriname increased sharply around 1941, in order to help build United States' military defense. The same year, the Paranam refinery situated near the Para river was put into operation.

In 1938, Billiton discovered high-grade, deep-seated bauxite in Onverdacht, Para district. As war broke out in Europe, and allied forces were short of bauxite, the government quickly arranged legislation to bring the abandoned Onverdacht plantation into development. This led to Billiton's first bauxite export of metallurgical, refractory and chemical grade bauxite in 1942 (Wong et al. 1998).

This helped Suriname to grow fast, and to become the world's largest bauxite/alumina exporter by the year 1947. Growth slowed down, in the period thereafter, due to development of bauxite industries in other countries in the region.

In 1965, Suralco started operation of their Paranam alumina refinery, in combination with construction of an aluminium smelter. Electrical power was supplied by the Brokopondo hydropower installation, built by the same company at Afobaka. This provided a boost to the export value of the country's bauxite resources for the following decades.

In addition, non-metallic grade bauxite was produced from Onverdacht and Moengo. In Moengo, Suralco was operating a calcining plant, with five drying kilns for the production of abrasive grade ore and a supercalcining plant at Paranam, for refractory grade ore. Billiton operated a dryer plant at Smalkalden (Bosma et al. 1973). In the 1960s and 70's, both companies produced about 130,000 tonnes chemical grade, 210,000 tonnes calcined grade and 35,000 tonnes of refractory grade ore per year, on average. Production of refractory grade bauxite ceased in 1983, and that of abrasive grade bauxite in 1987.

From 1986 onwards, all bauxite mined in Suriname was processed into alumina. In 1999, the smelter closed, while production and export of alumina continued until closure of the refinery in 2015. Bauxite production, for the best part of the 21st century, averaged about 5Mt per annum, with a bauxite blend from Moengo and Onverdacht areas which was refined to about 2.2 million tonne of alumina for export per year, with bauxite from Onverdacht generally characterised as high overburden/high AA (~50%)/low iron/high silica and from Moengo as lower AA (~44%)/high iron/low silica.

6.2.1 Bakhuis

For many decades past, the Bakhuis bauxite deposit has been considered as representing the long-term future of Suriname's bauxite industry. The Bakhuis Mountains form an elongated horst structure 25 km wide and 95 km long, with hills and plateaus at elevations ranging from 200 to 550 m. The area, which is underlain by Precambrian metamorphic gneisses and gabbros intrusions, has attracted interest since the early

1950s for lateritic nickel, manganese and bauxite. The first indication of bauxite was reported in 1957, and was followed by grassroots exploration in the early 1960s.

The Geological Survey of Suriname explored the whole Bakhuis region from 1961 to 1963. The reconnaissance methods included geological mapping and surface sampling. Of a total surface of 2400, 2000 km^2 were found to be prospective for bauxite. Surface sampling indicated that bauxite was approximately equally distributed throughout the structure. Ten prospective areas were selected for systematic drilling because of their accessibility. The drilling method was manual.

Four expeditions were carried out, from 1965 to 1966, by a consortium including Alcan, Alcoa, Billiton, Olin, and Revere. Drilling was performed by portable auger drill (Suralco) and Empire drilling (Billiton). A total of 65 holes in Area 1, 75 holes in Area 8, and 106 holes in Area 10-1 were drilled on a grid of 200 m × 200 m. Samples were collected at metric intervals.

From 1967, several plateaus on the western side were drilled by a joint venture between Kaiser and Pechiney, followed by an exploration program carried out by Reynolds and Grassalco (Grasshopper Aluminium Company), which is a state-owned company. This program took place in the period 1971–1975, and covered several plateaus in the north-east of the Bakhuis Mountains. In 1975, Reynolds pulled out of the project, which left the Bakhuis concession in the care of Grassalco. In support of the development of bauxite resources in Bakhuis, as a long-term option for Suriname's bauxite industry, a 72 km-long railway line was constructed during the period of 1976–1978, from Bakhuis' most northern plateaus (containing the highest grades) to the town of Apoera, on the banks of the Korantijn River (see Fig. 6.1).

A port facility and a rail maintenance facility were also built in Apoera, with accommodation for a future workforce, and some rail equipment was acquired as part of this project, although to no immediate effect (BHP Billiton reports, 2008).

A number of project evaluations, based on the Grassalco database, were conducted until 1997, including the feasibility study of 1979 by Billiton, that of 1980 by a joint commission (Government, Suralco and Billiton), and 1982 by Suralco, as well as several prefeasibility and opportunity studies. Two feasibility studies were carried out: one by Mackay and Schnellmann in 1997, and most recently by BHP Billiton Maatschappij Suriname, in 2008.

BMS' latest effort was close to execution, but failed due to non-technical reasons. About 7,500 holes were drilled—bringing the total number of holes drilled over the several programs combined to 13,000 holes—while over 90 sampling trenches were excavated between 2003 and 2007, as part of the associated exploration program, covering a permit area of 2,784 km^2.

Production of bauxite from West-Suriname has not yet commenced, despite the large investment made in exploration, studies and infrastructure. In Sect. 6.3, future options for Bakhuis will be discussed.

Fig. 6.1 Apoera to Bakhuis railway line (personal archive)

6.3 Future Options

The closure of the country's refinery at Paranam in 2015 has changed the bauxite business case, but undeveloped resources remain. Unlike gold, where potential for one or more yet undiscovered world-class deposits is present, Suriname's current bauxite potential is quite well defined. As described in Chap. 3, Suralco and Billiton Maatschappij Suriname (BMS) have surveyed and drilled the bauxite occurrences in the country as part of a series of studies intended to sustain their operations.

The remaining deposits are at Nassau Mountains, Lely Mountains, Browns-berg and Bakhuis Mountains, plus remnants existing around Onverdacht (mainly Para North/Kankantrie North) and Moengo (Coermotibo Deep-seated), as shown in Fig. 6.2. These were drilled in the period when producing areas around Onverdacht and Moengo were running out of bauxite supply, in order to assess the potential for the next generation of bauxite mines for the country.

Basic estimates of key parameters for the above resources are listed in Table 6.1.

The largest deposit is located in the Bakhuis Mountains, which contain more than 300 Mt of bauxite at a grade of 39% available alumina (or 44% total alumina), with possible extension into the Adampada-Kabalebo area. The proximity to the Central Suriname Nature Reserve in relation to plateau drainage areas will have an impact on future reserve estimations.

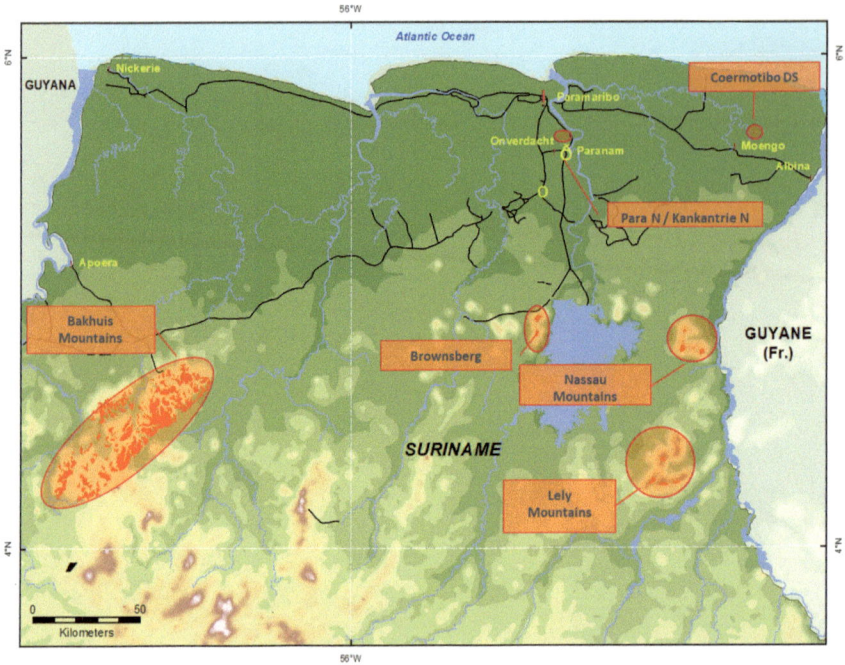

Fig. 6.2 Known and explored bauxite deposits

Table 6.1 Resource estimates of known bauxite deposits at 30% AA cut-off grade (BHP Billiton data)

Location	Tonnage (Mt)	%AA	%RSiO2
Bakhuis mountains	>300	39	2
Coermotibo deepseated	48	41	7.3
Nassau mountains	35	40	2.2
Para North/Kankantrie N	15	50	5.7
Lely mountains	13	36	1.9
Brownsberg	11	33	3

The business case developed for Bakhuis by Billiton Maatschappij Suriname, in the period 2006–2008, was based on transporting bauxite by rail to Apoera (a distance of 72 km), by barge via the Korantijn River, and along the coast to Paranam refinery, and finally export of alumina. Rail infrastructure from Bakhuis to the village of Apoera, on the banks of the Marowijne River, was already in place, but would have had to be rehabilitated, due to the long period of zero maintenance since its construction in the 1970s.

Fig. 6.3 Ten-year spot price for aluminium metal (with January data points from Indexmundi)

6.3.1 Bauxite Export

The market for export bauxite is driven by demand in China. Cost and grades of Al_2O_3 and SiO_2 are important indicators for relative competitiveness. Given the parameters listed in Table 6.1, Suriname's bauxite will not be in a strong position, compared to important competitors from Guinea and Australia, as discussed below.

The aluminium spot price averaged about 0.84 \$/lb or 1,850 \$/t in the period 2014–2018, as shown in Fig. 6.3. Bauxite prices are not (as yet) determined on an open spot market but are negotiated between suppliers and refineries, depending on product qualities.

6.3.2 Guinea

In the Atlantic market, Suriname would have to compete with Guinea, where massive bauxite deposits (over 40 billion tonnes in total) are present in flat topped plateaus at 200–400 m above sea level, with negligible overburden and topsoil cover. Alumina content varies from medium (40%) to as high as 52% in the lateritic bauxite deposits. On average, 46% total alumina (with about 42% available alumina) can be exported from these deposits, with easy access to road, rail, river and port infrastructure.

Guinean bauxite, especially from the Boke belt, has a fairly low silica content. About 50% of total silica is in the form of reactive silica, ranging from 1.2 to 1.8%, making Guinean bauxite ideal for processing in low-temperature refineries at 143 °C (Nandi, blog 2017).

Compagnie Bauxite de Guinee (CBG) has been a major exporter of bauxite, with current production rates of about 15 Mtpa, and currently expanding to 25 Mtpa. New suppliers are coming on-line rapidly, notably Societé Minière de Boké (SMB), a consortium established in 2014, including the Chinese Wei Qiao Group, the biggest aluminium manufacturer in the world. Their mine was fast-tracked to production within two years, and is serving Chinese alumina refineries. This company has built a mine with associated infrastructure, including four barge loading quays within the span of a year, and it uses a simple truck and barge operation to produce 15 million tonnes per annum, which is expected to rapidly increase to 30 Mtpa in the near future. Total bauxite export from Guinee has increased to 45 Mtpa in 2019, and experts expect that Guinee will overtake Australia as number one bauxite exporter.

6.3.3 Australia

Australia is currently the world's largest producer of bauxite, at 83 Mtpa, representing 28% of global production in 2017. The large bauxite resources at Weipa, with more than 3 billion tonnes in Queensland and Gove (>200 Mt) in the Northern Territory, have average grades between 49 and 53% total Al_2O_3, and are amongst the world's highest grade deposits located relatively close to the Chinese market. Other large deposits (each >500 Mt) are located in Western Australia (WA). The bauxite mines in the Darling Range of WA have the world's lowest grade bauxite mined on a commercial scale (around 28–33% total Al_2O_3). Despite the low grade of the ore, the mines accounted for 23% of global alumina production, as they also have low reactive silica, making the bauxite relatively easy to refine. Most bauxite is refined locally, but direct export of bauxite from WA to China has commenced.

6.3.4 Suriname

For direct bauxite export, the available grades at Bakhuis appear to be too low, (especially in comparison with existing competition), with about 80 million tonne at 45% available alumina at the tail end of the grade tonnage curve (see Fig. 6.4).

Grades at Nassau Mountains would require a blend with a higher-grade resource, like Para North and/or Kankantrie North, to create an export-quality blend. Total available tonnage at Nassau is limited, and it also requires an investment in road infrastructure, in order to transport bauxite to a loading port, which would require the equivalent of a new road of 140 km, in the case of shipping from Paranam.

Para and Kankantrie mines were operated by BMS until the stripping ratio made continuation uneconomic at the time. Resources extending northwards remain, however, at depths of >30 m. For development, a removal of significant volumes of overburden partly made up of soft demerara clays would be required before bauxite can be accessed. These volumes are estimated to be about 112 million m^3 including 50% soft, demerara clays. Besides significant development costs, the soft overburden poses geotechnical challenges, exacerbated by the fact that the area has been used to dump old dredge spoils. On the positive side, thanks to many years of continuous operation in the area, road infrastructure will require limited investment.

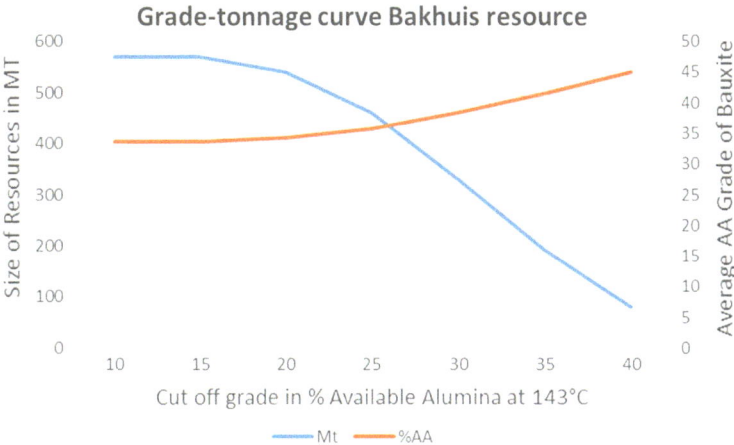

Fig. 6.4 Bakhuis' unconstrained grade tonnage curve (modified with data from BHP Billiton)

Coermotibo's deep-seated deposit contains very high grades of SO_3 (6%) which may require special treatment or washing. For development of this resource, the Coermotibo River will have to be diverted before mining can take place.

Brownsberg is currently a nature reserve, and therefore mining may not be allowed. The tonnage and grade of this resource do not justify environmental sacrifices.

The resource at Lely mountains, 50 km south of Nassau has lower grades than its neighbor, and will require significant investment in additional infrastructure (100 million US$) to make the area accessible.

6.3.5   Beneficiation

In 2006, BHP Billiton and Suralco looked at beneficiation options to possibly upgrade the AA-grade of Bakhuis, Nassau and Coermotibo deep-seated. In a beneficiation plant, Fe_2O_3 is rejected by a gravity process, which improves the alumina grade at the expense of volume.

Initial tests were carried out, to study the effect of such process using sample material from these sites. In summary: for all three locations, an upgrade of Available Alumina grade of 125% was achieved. The mass recoveries for Bakhuis and Nassau were 75% and 65%, respectively (Source: author's personal file).

The implication from above tests is that, in the case of Bakhuis, the sample grade of 38.5% was improved to 48% AA, with a resource ¾ of the original size.

The cost of a beneficiation plan was conservatively estimated at 150 million US$ at the time. For Nassau and/or Coermotibo deep-seated, the revenues over the life of the mine will be too small to write off investments of several hundred million in infrastructure and process, given the expected market price.

6.3.6 Alumina

For treatment of metallic grade bauxite to alumina, only Bakhuis has the size and potential to invest in a (small) refinery. A new refinery would require access to (cheap) energy. Heat is required for the refining process, and could be produced by Staatsolie's fuel oil or off-shore gas. A location near Nickerie would be an option, with the additional benefit of direct access to the Atlantic Ocean.

The recent oil discoveries in the Liza field of the Stabroek block, 200 km off the coast of Guyana, are an interesting development (see Fig. 10.1). In 2016, Exxon confirmed a world-class resource discovery, in excess of 1 billion oil-equivalent barrels. In 2019, the company has signed an order for a multi-purpose hull designed to produce 220,000 barrels of oil per day, along with a gas treatment facility (www. sbmoffshore.com 2019). The year 2020 started with an announcement by a consortium of Apache and Total that oil was discovered at the Maka Central-1 exploration well, 150 km offshore in Suriname's own coastal waters. This could offer even greater opportunities for vertical integration in the country.

The most recent feasibility study on the Bakhuis project was carried out by BHP Billiton, in 2008, and was based on a mine with a production capacity of 2.5 million dry tonnes per year, rehabilitation of 72 km railway to Apoera, construction of a loading facility at Apoera to load shallow-seagoing barges, and an off-loading facility at Paranam. The total project estimate was about 730 million US$.

In order to achieve a rough estimate for a similar case—with rail transport to a new refinery located at Nickerie—the following changes to the original formula are made:

- Add 25% inflation
- Remove costs related to barge operations
- Extend rail by 90 kms.

This brings a new estimate to about 800 million US$, excluding refinery costs.

The current estimated cost to build a small alumina refinery in India is around 650 US$ per ton alumina produced, and about 600 US$ for one in China, excluding the cost to build a power plant (Nandi, personal communication). These estimates would have to be escalated for application in Suriname, in order to allow for additional freight costs.

Alumina is currently trading at 380 $/t. The price has improved significantly since leading alumina producers managed to step away from the contract system, which linked the alumina price to the aluminium price. The alumina price index (API) was created in order to reflect its own cost base and supply/demand fundamentals (Fitzgerald 2019).

6.3.7 Non-metallic Bauxite

Guyana and China are the world's major producers of these special types of bauxite.

Refractory grade bauxite typically has very high alumina grade (minimum 59–61%), with grades of SiO2 maximum 1.5–5.5%, Fe_2O_3 maximum 2% and TiO_2

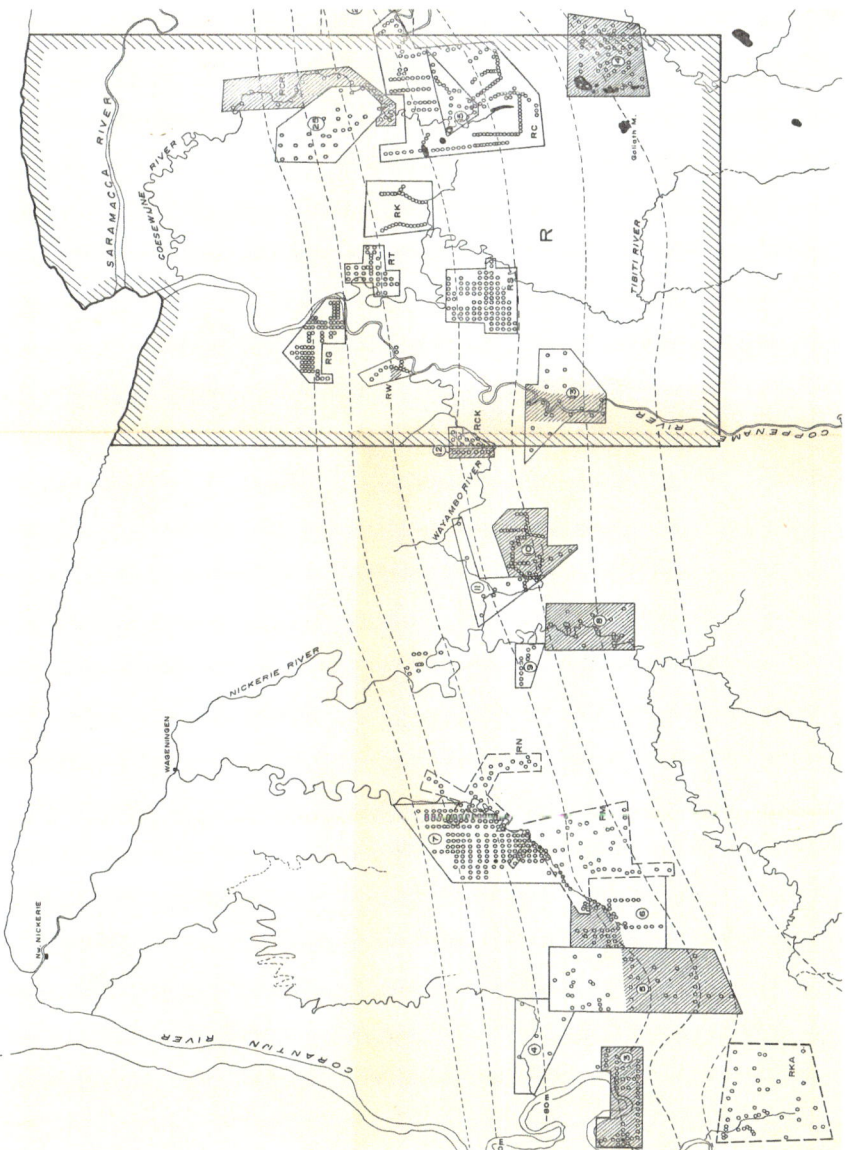

Fig. 6.5 Map with exploration permits drilled in the coastal plains of West-Suriname. For scale: the distance between Nw. Nickerie and the mouth of the Saramacca River is 120 km (Van Lissa 1975)

maximum 2.5% (Hill and Sehnke, SME 2006). Thanks to higher revenues per ton, much smaller tonnages are required for economic exploitation. Prices depend on product type and quality; they are usually based on fixed supply contracts. Suriname was an exporter of special grade bauxite until its supply ran out, but there remains potential for unexplored resources.

As stated in Chap. 3.2, selected areas between Saramacca and Korantijn Rivers were explored for bauxite in the 1950s by Reynolds Metals Company and Guyana Exploration Company, who acquired a number of concessions distributed over the coastal plain. Both companies drilled more than 1,000 exploration holes in these areas. In 1975, Van Lissa reviewed the exploration work, and concluded that in several blocks (specifically numbers 4, 5, 9, 13 and 25 as shown in Fig. 6.5), holes reached maximum depths in the range of 22–39 m, and did not reach basement. This suggests that deeper bauxite could have been missed, as illustrated by the relevant example of Tarakuli in Guyana, which is a deposit located 15 km inland across the Korantijn River.

Reynolds' drill data from Tarakuli in the 1960s indicated an historical, inferred bauxite resource of approximately 62.7 million MT at 58.6% Al2O3, 4.7% SiO2, 2.5% TiO2 and 3.3% Fe2O3. Half of the tonnage was classified as metallurgical grade bauxite, and half as chemical grade bauxite. The average ore thickness is 6.6 m, and it is located below an average overburden cover of 46 m (First Bauxite 2010).

This deposit sits well below the average depths drilled in several blocks, as specified before (and in theory), indicating that an occurrence at similar depth could exist on the Suriname side.

References

Bosma W, Ho Len Fat AG, Welter CC (1973) Minerals and Mining in Suriname-Mededelingen Geologisch Mijnbouwkundige Dienst Suriname 22:71–101

First Bauxite Corporation (2010) 21/7/2010 News announcement. https://www.firstbauxite.com

Fitzgerald B (2019) Pricing revolution to underpin alumina expansion for Alcoa/Alumina—MiningNews.net

Hill VG, Sehnke ED (2006) Bauxite—Industrial minerals and rocks industrial minerals. In: Kogel JE, Trivedi NC, Barker JM, Krukowski ST (eds) SME, 7th edn, pp. 227–261

Monsels D (2016) Bauxite deposits in Suriname: geological context and resource development—Netherlands. J Geosci 95(4), 405–418

Nandi AK (2017) Why is Guinea bauxite considered the best in the world? https://www.blog.alcircle.com

Van Kersen JF (1956) Bauxite deposits in Surinam and Demerara (British Guiana)—Thesis Leiden. Also published (1956) in Leidse Geologische Mededelingen, vol 21, pp 247–375

Van Lissa RV (1975) Review of bauxite exploration in the coastal plain of Suriname. Mededeling Geologisch Mijnbouwkundige Dienst 23:250–254

Wong TE, de Vletter DR, Krook L, Zonneveld JIS, van Loon AJ (eds) (1998) The history of earth sciences in Suriname, 479 pp

Chapter 7
Base Metals

Base metals are commonly defined as being metals of comparatively low value and as being relatively inferior in certain properties, such as corrosion. They are, however, endowed with many unique and important properties. Besides aluminium, this group of metals includes copper, lead, zinc and nickel.

7.1 Copper

Copper is an important metal in the modern world, and its demand will continue to grow, driven by both urbanization and energy transition. An electric car uses four times more copper than conventional gasoline-powered vehicles, with wind and solar energy expected to drive future copper demand even more. Power generation accounts for almost half of China's copper use. The World Bank estimates a cumulative demand of 20Mt copper metal until 2050, to meet energy transition goals as defined in the Paris Agreement of 2016 (World Bank 2017).

Not enough copper is being discovered to meet future projected demand (S&P Global Market Intelligence), with a 5.7 million tonnes deficit projected for 2030. Copper prices are expected to increase to 10,000 $/tonne in coming years (Kitco News, Feb 2019).

7.1.1 Exploration Results

The GMD has located copper-bearing laterite on top of a small hill of about 1 km^2 located at a distance of 10 km west of Weko Soela and 40 km southwest of Palumeu, in the area west of the Tapanahony river. Geochemical sampling and some shallow drilling to bedrock was carried out, between 1960 and 1967. The deposit location is shown in Fig. 7.1.

© The Author(s), under exclusive license to Springer Nature Switzerland AG 2020   43
M. Keersemaker, *Suriname Revisited: Economic Potential of its Mineral Resources*,
SpringerBriefs in Earth Sciences, https://doi.org/10.1007/978-3-030-40268-6_7

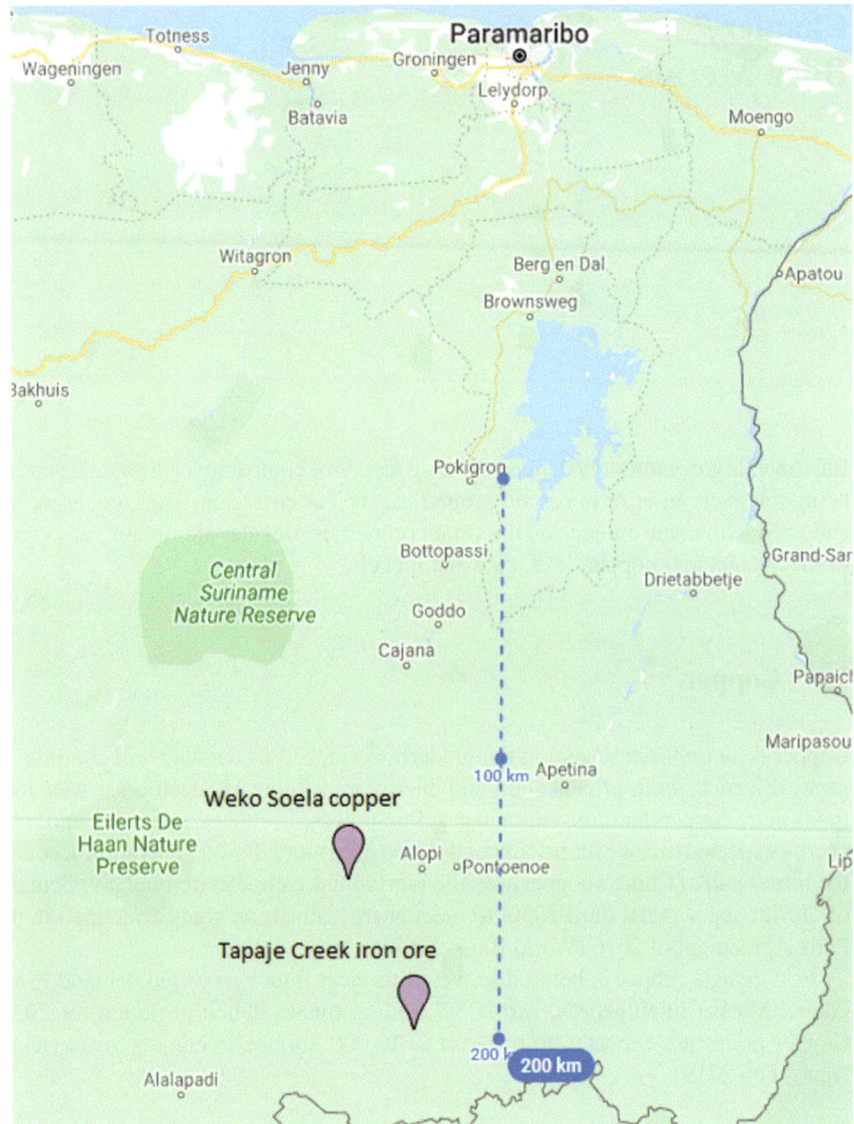

Fig. 7.1 Map of Suriname with locations of Weko Soela copper, Tapaje Creek itabiritic iron ore and Pokigron at the south end of existing road infrastructure network

The deposit at Weko Soela contained 0.3–0.8% copper in the laterites, included values of 0.8 and 4.5% copper in the weathered rock, and up to 0.5% copper in bedrock, with turquois as main copper mineral in weathered rock, and laterites, chalcopyrite and bornite in the bedrock. The host rock is an unusual igneous cordierite diorite, probably consisting of a smelt from migmatitic politic gneisses. The copper minerals occur together with magnetite and apatite as intercumulus minerals between a cumulate of euhydral cordierite crystals (Kroonenberg 1977 and personal communication).

In 1969, Suralco obtained an exploration license covering an area of some 20,000 ha in the area around Weko Soela. A program of geochemical work and an IP survey were carried out. Based on preliminary field results, a drilling program followed in anomalous areas in Weko Soela and in the nearby Awalape Creek. In all, 7 holes or 232 m were drilled, including 90 m core. Five holes reached depths between 24 and 33 m with two more of 81 m and 52 m depths. Highest grade sampled was 0.5% from hole 6 at Weko Soela area (5010 ppm between -57.0 and -57.1 m) (Keyzer 1972; Wadell 1972). It was noted that one anomaly in area D was not drilled, due to logistical problems.

Exploration has been carried out for copper in an area of about 5 km^2 around Wedeboh Soela, along the Saramacca River. The area consists of granitic rocks cut by a few dolerite dikes and quartz veins. Locally, minor amounts of copper minerals are present in these rocks: chalcopyrite and bornite in granite, digenite and brochantite in a quartz vein at Wedeboh Soela, and native copper in the dolerite dikes (Bosma et al. 1973; Wong et al. 1998).

Basic rocks of the more volcanic part of the Marowijne Group locally contain minor amounts of copper sulphides, particularly chalcopyrite, for instance near Jandee Creek near Rosebel; the same applies to certain dolerites and gabbros (Doeve 1966).

Metallic copper would be present in alluvial deposits near Benzdorp. Green oxidic copper minerals have been reported in an area near the Oelemarie airstrip, in the southeast of the country (Bosma et al. 1973).

Another copper occurrence was found in the Bakhuis Mountains along the Upper Nickerie River, as indicated in Fig. 7.2.

This occurrence has been fairly thoroughly investigated, including 63 vertical diamond drill holes over a strike length of 2.5 km (Dahlberg 1982). The drill core with the strongest copper mineralised rocks averaged 0.33%, including a 22 m intersection assaying 0.25% Cu and about 10 m predominantly monzonitic rock assaying 0.69%. The best two-meter interval of drill core assayed 1.56%. The holes did not exceed -100 m. The copper minerals are bornite and chalcopyrite. The copper occurrence was evaluated by the UN Revolving Fund for Natural Resources Exploration, and was not considered worth further exploration (Wong et al. 1998).

It is interesting that—partly coinciding with the copper anomaly—a 1,500 m long zone of phosphate enrichment was found in the soil, caused by sub-vertical, apatite lenses of up to 12 m in length, as further described in Sect. 9.5. In addition, anomalous values were found, consisting of thorium and also of the Rare Earth Elements cerium, lanthanum, yttrium, etc., which were probably partly derived from monazite (De Vletter 1984). There is no direct geochemical or petrographic link between the copper and phosphate mineralization (Patadien et al. 2019).

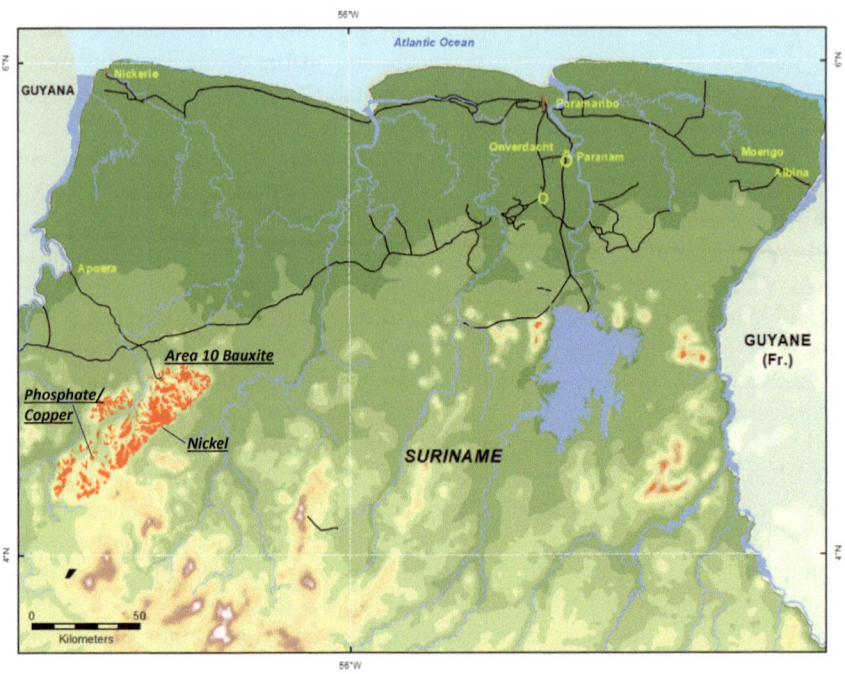

Fig. 7.2 Copper, phosphate and nickel deposits at Bakhuis mountains

7.1.2 *Economic Evaluation*

More than half of the world's copper production is supplied from porphyry copper deposits, with low copper grades generally between 0.5 and 1.0%, and very large volumes. Most are relatively near the surface, lending themselves to large-scale bulk mining at low cost, as the entire body of shattered rocks with tiny veinlets is mined. Chile hosts some large mines, such as Escondida, the world's largest copper mine, operating at a remaining copper reserve grade of 0.52% (Visualcapitalist.com).

A relevant deposit for comparison with Suriname is the Salobo mine in northern Brazil, which is the largest copper deposit ever discovered in Brazil, and which is now producing 250,000 tonnes of copper per year, from a resource of a billion tonnes at 0.85% Cu. This deposit was discovered only fairly recent, in 1977, and is an example of an iron oxide copper gold (IOCG) deposit with biotite and magnetite schists as its major hosts. The large fraction of iron oxides in the ore makes this type a different class from porphyries (Pohl 2016). Salobo has a massive magnetite body with small veins and irregular masses of biotite (Burns et al. 2019). The deposit is situated in an equivalent of the Guiana greenstone belt. Biotite and magnetite are also present in abundance in the Weko Soela deposit, with pyrite, chalcopyrite and bornite associated with magnetite in small amounts (Kroonenberg 1977).

Another significant type of copper deposits are the stratabound sediment-hosted deposits with copper grades up to several percent, in combination with cobalt as a by-product. Important examples occur in the African Copper Belt in Zambia and DRC, like the Kalongwe deposit used in the reference example below.

7.1.3   Kalongwe Copper Project

To investigate the minimum economic size of a copper deposit, Weko Soela is taken as possible location. For economic reference, the costs are compared with the Kalongwe Copper Cobalt project on the African Copper Belt. This is a low capex, low opex mining project, for which a feasibility study has just been completed, and which is about to be developed by its new Chinese owners. The project intends to produce 20,000 tonne Cu/Co in concentrate of 15% per annum from an open pit with a 1 Mtpa on-site dense media separation plant. The deposit contains a reserve of 8 Mt at 2.9% copper and 0.3% cobalt, and is expected to support a mine life of eight years. The mine is located 77 km away from a copper smelter (Nzuri Copper Limited website) (Table 7.1).

The capital investment is estimated at 53 million US$ and opex at 1.40 $/lb (corrected for cobalt credits, MK). Based on Kalongwe's estimates, the following additions were included for the Weko Soela case:

- An extension of 150 km to connect to the Afobakka road at Pokigron (see Fig. 7.1; estimated cost ~150 million US$)
- Road haulage equipment (~5 million US$)
- Removal of 30 m overburden, of which 50% or 15 million bank m^3 in pre-development @ an estimated 5 $/m^3 (~75 million US$).

Total investment required is rounded to 300 million US$.

Table 7.1   Kalongwe project parameters (Nzuricopper.com)

Stage 1-Financial results	2018 Updated FS
NPV$_{10\%}$ US$ (pre/post-tax)[a]	US$186 M/US$130 M
IRR % (pre/post-tax)[a]	99%/76%
Annual average production (cu/co-in-concentrate)	18,657 t cu & 1,370 t co
Total production LOM (cu/co-in-concentrate)	149,258 t cu & 10,964 t co
LOM (years at 1Mtpa throughput)	8 years
c1 cash cost US$ (including co credits)	US$0.85/lb
CAPEX US$ (excluding working capital, ±15% accuracy)	US$53.12 M
Payback (months)	17 months

[a]NPV/IRR based on US$3.00/lb LME cobalt sales price and a 100% project basis. The proposed 2018 DRC mining code changes are not included nor considered to apply to the Kalongwe Project at this time

Additional operating costs include:

- 300 km haul by road to Paranam Port of 400 tonne concentrate per day
- Shipment by sea to the nearest copper smelter in Bahia, Brazil by Handysize charter (0.06 $/lb).
- Stripping costs for the remaining 15 million m³ of overburden at 0.2 $/lb.

The copper price is based on the average proce over the last 10 years as shown in Fig. 7.3. Applying the same formula used in for gold, as outlined in Chap. 5, the following results are obtained. The recovery factor was copied from Nzuri's feasibility value of 63.5% (Table 7.2).

The above calculations suggest that a minimum of 536 kt copper has to be present, which is equivalent to 18 Mt with a grade of about 3% to make an economic case for exploitation. The copper occurrences identified to date all are fairly remote, and would have similar threshold values.

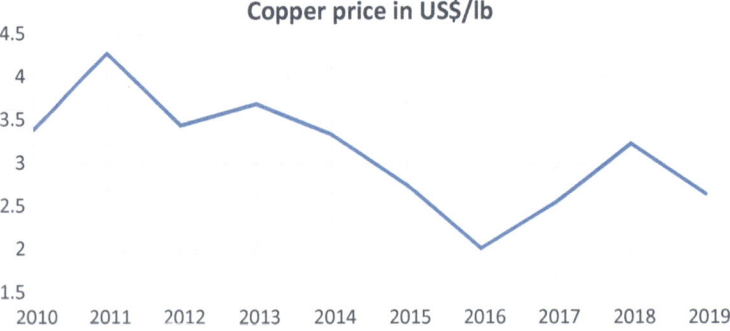

Fig. 7.3 Ten-year copper price in US$/lb (with January data points from Macrotrends.com)

Table 7.2 Proven and probable copper reserves required at Weko Soela

P + P formula			
	Capital	300	MUS$
	Copper price	3	
Less	Opex	1.65	
		1.35	
Less	Royalty&tax	0.35	11.5% of cu price
		1.00	
Less	P + R	0.6	20% of cu price
Balance:	Capital/lb	0.40	
Minimum P + P		750.0	M lb
Contained copper (t)		340	'000 tonne
Contained copper (t)	at 63.5% recovery	536	'000 tonne

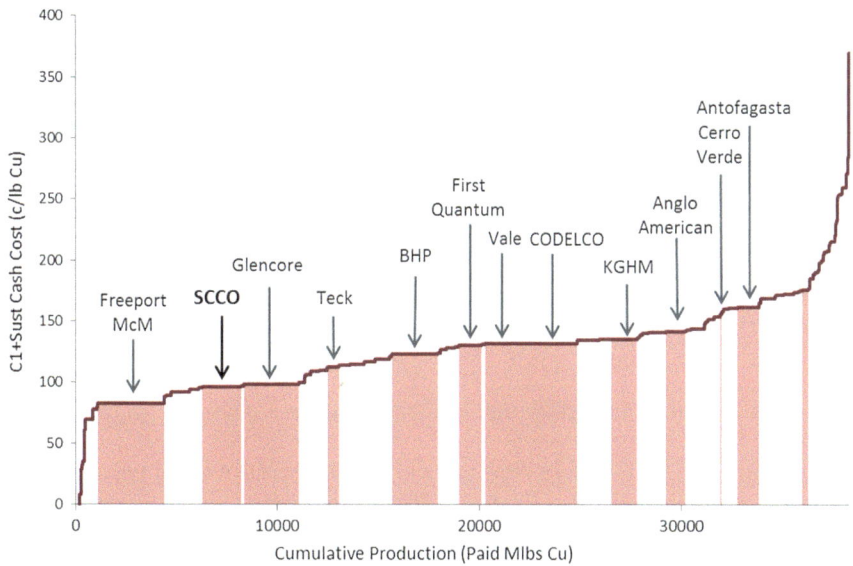

Fig. 7.4 Wood Mackenzie copper mine cost model 2017 (www.southerncoppercorp.com)

The business case would not alter significantly if Brazil becomes accessible by road, as has been suggested by President Bolsonaro. The distance from Weko Soela to the Amazon River is almost 600 km, in a direct line.

Another indicatoris the cash production cost per lb copper relative to world competition, as shown in the graph below, with data from 2017 (Fig. 7.4).

In 2017, the global weighted average production cash cost was 145.3 c/lb, strongly suggesting that (using the above projections) the Weko Soela project would be in the highest cost quartile. Salobo operates in the lowest quartile, so a similar business case could conceivably follow, in relation to a similar, massive deposit at Weko Soela.

7.2 Lead/Zinc

Lead and zinc are metals of comparatively low value and relatively inferior in certain properties, such as corrosion, but with some unique and important properties.

On the Guiana Shield, there is only one known volcanic-sedimentary base metal sulfide deposit (VMS deposit containing copper, lead and zinc). The Montagne d'Or deposit is a gold-bearing variant, and is located in French-Guyana (Guiraud et al. 2017). It has been debated whether this is due to a veritable deficiency in these metals, or to the difficulties of finding them because of deep lateritisation or other known factors, such as dense forest and a lack of infrastructure (de Vletter 1984). As mentioned in Sect. 7.1, the discovery of Salobo deposit supports the views of optimists that more deposits are still hidden.

The comparable Birim greenstone belt hosts the Perkoa underground zinc mine in Burkina Faso (see Fig. 5.3), with a measured and indicated resource of 3.1 Mt at 12.7% Zn (www.trevali.com, 2019). This mine sits, in ranking, just outside the top ten zinc mines in the world, in terms of output per annum, at 80,000 tonnes. This volcanogenic massive sulfide deposit is the only one of its kind in the Birimian Supergroup of West Africa. The ore is mainly composed of Fe-rich sphalerite (30%), pyrite (25%) and barite (10%) (Schwartz 2003).

7.2.1 Exploration Results

Doeve reported traces of sphalerites in the pegmatites around Jorka Creek (Doeve 1966).

From a period beginning in 1969, soil samples were collected in a regional recon-naissance program covering an area of 670 km^2 south of Moengo, on the so-called pegmatite belt (Welter 1975).

A total of 1,145 samples were collected along sets of parallel lines running 2.5 km apart. Besides revealing high values for tin, relatively high values Zn concentrations were found near contacts of granites and metasediments.

Following up on Doeve's work, many locations were included in a country-wide geochemical program, where Pb and Zn were included in standard analyses, as described in Sect. 3.1 and shown in Fig. 3.1 (Oosterbaan 1973) and as described by Verwey (1984).

7.2.2 Economic Evaluation

Glencore invested 220 M US$ to develop Perkoa, in 2013. The milling and process-ing plant, including dense media separation and flotation cells, combined with the underground mine facility to require significant investments; hence the need for a large-scale operation and reserves of 873,400t of contained zinc metal, at a cut-off rate of nine percent, in this case (mining-technology.com). In 2019, zinc generates about 1.20 US$/lb.

7.3 Nickel

About 65% of the nickel consumed in the Western World is used to make austenitic stainless steel. Another 12% goes into super-alloys or nonferrous alloys, and 3% goes into batteries used in electric vehicles and in renewable energy storage

In recent years, production of refined nickel decreased, as stainless steel producers, primarily in Asia, preferred lower cost nickel pig iron. This is low-grade ferronickel, which is a cheaper alternative to pure nickel, and which contains iron as a bonus. Mine production in countries that supply direct shipping ore to nickel pig iron operations increased, while mine production supplying refineries tended to decrease. Production of nickel chemicals, however, has increased, particularly in the case of nickel sulfate, which is used in the production of batteries. Industry analysts project a significant increase in global nickel consumption in batteries for energy storage and electric vehicles (USGS, 2019).

Increase of batteries used in electric vehicles is growing at almost 16% per year, driven by strong policy stimulus from Chinese and Western governments. Expectation is that projected supply additions of nickel will fall short of projected demand by 2025. This will have a positive effect on the price of nickel (McKinsey 2018).

Two main types of nickel deposit are currently exploited worldwide: sulfide (in which nickel is associated with copper and platinum-group elements) and laterite (in which nickel is associated with cobalt) (Evans et al. 2016).

7.3.1   Exploration Results

Exploration for nickel was carried out on a small (2×1.5 km^2) and low grade lateritic nickel occurrence, overlying basic to ultrabasic rocks in the Adampada area, in the southeast part of the Bakhuis Mountains, about 8 km north of the Left Adampada Creek, and about 21 km south of the railhead at Area 10 (see Fig. 7.2). The area is accessible via the Bakhuis road network developed for the bauxite project. The parent rocks consist, for the greater part, of (meta)-gabbros, but in one area the rocks are mainly ultrabasic. These rocks are considered to belong to the De Goeje Gabbros (Wong et al. 1998).

A total of 18 diamond holes were drilled, in an area of 1 km^2 to a maximum depth of 175 m in the centre; other holes were drilled to various depths less than 100 m. The chromium content varies between 0.15 and 0.20%; cobalt values are up to 350 ppm (equals 0.035%), and copper less than 100 ppm. Sulphide minerals were not seen in the cores (Oosterbaan 1975).

Between 1969 and 1971, 65 auger holes were drilled in an area of 6 km^2, thus forming part of an anomaly. Samples were taken at every meter. The zone richest in nickel varies from 1 to 6 m in thickness, at depths ranging from 5 to 20 m within the weathering profile. An exceptional (chlorite-rich) sample assayed 2% Ni.

From one-third of all drill holes, samples were analysed for copper. Values of about 100 ppm were found all over the profile, with local concentrations up to 750 ppm. Chromium contents determined in a limited number of samples are around 0.5%, and locally up to nearly 2%. From a number of samples with high Ni values, Cobalt was analysed, ranging from 0.1 to 0.12%, averaging at about 0.04%.

The table below gives Oosterbaan's estimated resources for various cut-off grades, assuming a density of 1.5 t/m^3 (Table 7.3).

Nickel-bearing talc schists were found in the Bemau Creek area 6 km west of Lemmiki, on the Saramacca River (Beckering Vinckers 1958). The ultrabasic rocks

Table 7.3 Resource of lateritic nickel as per Oosterbaan (1975)

Cut-off grade (%)	Area (km2)	Thickness (m)	Tonnage (million tonnes)	Average grade (%)
0.6	0.92	2.8	3.9	0.81
0.75	0.8	2	2.4	0.93
1	0.25	1.75	0.7	1.22

in the area form a ridge with an extension of about 25 km². Diamond drilling was done at various locations, partly coinciding with anomalies. The ultrabasic rocks proved to contain Ni generally less than 0.3%, with Cr (Chromium) values of the same order and a few Cu values up to 0.2%.

The ultrabasic rocks are covered with a lateritic soil of varying thickness (maximum 47 m), which contain up to about 0.25% Ni. In the lateritic overburden, one of the drill holes near Dramhoso intersected a corundum-spinel rock, with local values of 0.7% Ni and 6.5% Cr (Bosma et al. 1973; Veenstra 1978).

7.3.2   *Economic Evaluation*

Nickel price has experienced some aggressive increases and drops in the past two decades, with lows at 5,000 US$/t in 1999 rising to a peak of 54,000 US$/t in 2007. Current prices are around 14,000 US$/t as shown in Fig. 7.5. The nickel sector is experiencing new mine project developments following a calm period after the storm, during which time no investments were made.

Fig. 7.5 Ten-year nickel price (with January data points from indexmundi)

Table 7.4 Project parameters for Barnes Hill, Tasmania

Barnes Hill Project	Base
Capital cost ($ millions)	78.4
Project ROM tonne throughput (million t)	500,000
Nickel grade first 5 years (%)	1.01%
Nickel grade second 5 years (%)	0.73%
Nickel recovery (%)	90%
Life of mine nickel price US$/lb	10.00

7.3.3 Nickel Laterite

Typical nickel laterite ore deposits are very large-tonnage, low-grade deposits located close to the surface. They are typically in the range of 20 million metric tonnes and upwards, with grades around 1–1.5% Ni. They are growing to become the most important source of nickel metal in relation to world demand (currently second to sulfide nickel ore deposits).

New Caledonia contains 21% of the world's nickel laterites, followed by Australia (20%), the Philippines (17%), and Indonesia (12%).

The list of "New Nickel laterite projects by year end 2017," published by Index Mundi in 2018, includes a range of Nickel laterite development projects, with grades ranging from 0.6–2.1% and resource tonnage estimates ranging from 2.5 to 490 Mt.

At the lower end of the scale sits the Barnes Hill project on Tasmania, with cobalt credits and situated 15 km from a deep-water port. Their resources contain 2.4 Mt Ni of 0.94% and 0.06% Co, at a cut-off grade of 0.8% Ni (USGS and 99mines.com).

The owner is completing a feasibility study on a cut-off of 0.5% Ni, for a resource of 6.6 Mt at 0.81% Ni and 0.05% Co. The plan is to produce nickel cathodes with below-base parameters (pamcoal.com), including an optimistic view which projects nickel price to rise to well above today's price (Table 7.4).

Compared against the above yardstick, Bakhuis' Nickel laterite project—as estimated by Oosterbaan in 1975—would be an average-grade but small project, and would certainly be worth exploring further, when plans to develop Bakhuis infrastructure for bauxite (and phosphate) exploitation are reconsidered.

References

Beckering Vinckers H (1958) Een orienterend onderzoek naar de aanwezigheid van nikkelerts in het gebied van de Bemaukreek. Geologisch Mijnbouwkundige Dienst Suriname

Bosma W, Ho Len Fat AG, Welter CC (1973) Minerals and Mining in Suriname. Mededelingen Geologisch Mijnbouwkundige Dienst Suriname 22:71–101

Burns N, Davis C, Diedrich C, Tassami M (2019) Salobo copper-gold mine Carajas, Para State, Brazil. Wheaton Precious metals technical report

Dahlberg EH (1982) Geochemical investigation of magnetic and electromagnetic anomalies in the Upper Nickerie copper-rare earth mineralization area, Suriname, pp 95–109. In: Laming DJC, Gibbs AK (eds) Mineral exploration techniques in tropical rain forest–AGID Report 7

De Vletter DR (1984) Economic geology and mineral potential of Suriname. Mededelingen Geologisch Mijnbouwkundige Dienst Suriname 27:11–30

Doeve G (1966) Delfstoffen in Suriname, mededeling 15 van Geologisch Mijnbouwkundige Dienst van Suriname 94–107

Evans DM, Simmonds J, Hunt JP (2016) An overview of Nickel Mineralisation in Africa with emphasis on the Mesoproterozoic East African Nickel Belt. Episodes 39:319–333

Guiraud J, Tremblay A, Jebrak M (2017) The Rhyacian "Montagne d'Or" auriferous volcanogenic massive sulphide deposit, French Guiana, South America: stratigraphy and geochronology. In: Conference: 14th SGA Biennial meeting, Québec, volume: proceedings, vol 1

Keyzer S (1972) Samenvatting van de Acker boorprestaties ivm Copper Exploration, Weko Soela— Gedurende de periode van November 1971 tot en met mei 1972. Geologisch Mijnbouwkundige Dienst Suriname

Kroonenberg SB (1977) Petrography of the copper-bearing rocks from Weko Soela, boven-Tapanahony, Suriname. Geologisch Mijnbouwkundige Dienst Suriname, Internal Report 7 pp

McKinsey (2018) Power play in mining magazine June 2018, pp 34–40

Nzuri Copper Limited: Project Overview Aug 2018. https://www.nzuricopper.com.au

Oosterbaan WE (1973) Geochemical exploration in Suriname—a review. Mededelingen Geologisch Mijnbouwkundige Dienst Suriname 22:1103–1118

Oosterbaan WE (1975) The Adampada lateritic nickel deposit. Mededelingen Geologisch Mijnbouwkundige Dienst Suriname 23:206–213

Patadien RS, LaPoint DJ, De Roever EDL (2019) The K3 copper deposit in the Bkahuis Granulite Belt, W Suriname. Mededelingen Geologisch Mijnbouwkundige Dienst Suriname 29:151–154

Pohl WL (2016) Economic Geology, Principles and Practice: Metals, Minerals, Coal and Hydrocarbons—an Introduction to Formation and Sustainable Exploitation of Mineral Deposits, John Wiley and Sons, pp. 712

Schwartz MO (2003) The Perkoa Zinc deposit, Burkina Faso. Econ Geol 98:1463–1485

Verwey R, De Vletter DR (1984) Economic Geology and mineral potential of Suriname. Mededelingen Geologisch Mijnbouwkundige Dienst Suriname, Geochemical Surveys 27:101–103

Veenstra E (1978) Petrology and geochemistry of sheet stonbroekoe—sheet 30, Suriname, Thesis Amsterdam, 161 pp. Also published (1983). Mededelingen Geologisch Mijnbouwkundige Dienst Suriname 26:138 pp

Wadell JS (1972) Final report weko soela-awalape copper prospect Upper Tapanahony River, Suriname. Geologisch Mijnbouwkundige Dienst Suriname/Suralco Interim and Final Reports

Welter CC (1975) Tin, copper and zinc in soils around Patamacca (NE Suriname). Mededelingen Geologisch Mijnbouwkundige Dienst Suriname 23:239–243

Wong TE, de Vletter DR, Krook L, Zonneveld JIS, van Loon AJ (eds) (1998) The history of earth sciences in Suriname, 479 pp

World Bank Group (2017) The growing role of minerals and metals for a low carbon future. World Bank Publications

Chapter 8
Other Metallic Minerals

8.1 Chromium

Chromium is essential to stainless steel production, thanks to its oxide-forming properties; there is no substitute for it. Among a variety of uses, chromium is a key component of certain widely-used alloys, e.g. aluminium alloys.

World resources are greater than 12 billion tonnes of shipping-grade chromite, an amount which is sufficient to meet conceivable demand for centuries to come. The world's chromium resources are heavily geographically concentrated (95%) in Kazakhstan and in the Bushveld complex of South Africa. The latter contains approximately 70% of the world's economic chrome ore reserves (www.samancor. com 2019).

For reference: beside a small occurrence of chromium in Sierra Leone, the Hangha deposit, there are no significant chromium deposits on the Man-Leo Shield.

8.1.1 Exploration Results

In 1971, a chromite occurrence was investigated near the Upper Saramacca River, 15 km east of the Tafelberg. Pitting and drilling was done to determine the size of the deposit, which led to an estimation of some 14,000 t chromite, with an average content of 50% Cr_2O_3 embedded in 24,500 m^3. This estimate was based on 13 surface pits of 18 m^3 each. The chromite bearing ultramafic rocks form an outlier on the Marowijne Greenstone Belt.

Five inclined holes were drilled in the deposit, to an average depth of 50 m, but they failed to show a massive deposit. The highest value was 14% Cr over 1.5 m, at a depth of 30 m. The average overburden thickness was about 20 m (Den Hengst 1975). The conclusion was that the potential for large resources in these chromium-bearing xenoliths is minimal.

M. Keersemaker, *Suriname Revisited: Economic Potential of its Mineral Resources*, SpringerBriefs in Earth Sciences, https://doi.org/10.1007/978-3-030-40268-6_8

A few years earlier, several small residual chromite deposits were investigated near the Emma Range, 25 km to the northwest at upper Toekoemoetoe Creek, lying just north of the Plet Ridge manganese deposit. They are connected with a 2 km-long zone of weathered chromite bearing ultrabasic rocks (Bosma et al. 1973; Wong et al. 1998). The weathering persists to depths of about 40 m (de Vletter 1984). After drilling and pitting, a resource was estimated of 154,000 tonnes of chromite, at about 30% Cr (based on random samples) in the upper 20 m below the surface, embedded in 1.4 million m^3 of weathered material (Bisschops 1969).

More chromite float was discovered later, in a small area west of the Upper Saramacca River (Den Hengst 1975). This occurrence was roughly estimated at 200,000 tons (Wong et al. 1998).

Similar chromite-bearing rocks were found at Pique Hill. The Cr content of rocks and stream samples ranged between 1–1.5%. Ultrabasic rocks from a few diamond drill holes west of Dramhoso, near the Upper Saramacca River, contained up to 6.5% Cr with 0.7% Ni. Minor amounts of chromite are widely distributed in alluvial deposits on both sides of the Fatoe Swietie Mountains east of Benzdorp. Nickel-bearing lateritic soil north of the Left Adampada Creek contained up to 3.0% Cr (Bosma et al. 1973).

8.1.2 Economic Evaluation

Unlike most other major metals that can be traded through metals markets, such as the London Metal Exchange, chromium–as either chromite or ferrochrome–is only bought and sold under contract, mostly an arrangement made directly between the miners and the foundries. As such, there is no avenue for speculators to influence the price, as generally happens for most metals.

The relatively constant price, from 2009 to date, reflects a balance of supply and demand made possible by the contracts between the miners and the foundries. The foundries pay enough to guarantee delivery, and supply shortages are rare. As such, there is very little opportunity for a potential new supplier to enter the market.

In this tight market, Suriname would have to compete with Samancor's operations on the Bushveld. At the end of June 2002, Samancor's proven reserves totaled 16.6 Mt grading 42.4% Cr_2O_3 with probable reserves of 23.4 Mt. Total resources are estimated to be sufficient for 200 years mining at current rates. The deposits in Suriname are thus too small and low-grade to become economically feasible.

8.2 Iron

8.2.1 Exploration Results

Iron is locally found in two types of deposits: itabirites and laterites. Itabirites or magnetite-hematite quartzites (Ornstein and Haug 1965) are found on plateaus near Tapaje Creek, just north of the border with Brazil, in the far south of the country, which form an enclave of Marowijne Greenstone Belt surrounded by Older Granites. Significant volumes have been identified at grades varying from medium (24–45%) to high-grade (60%) (Bosma et al. 1973).

Reserve estimates dating back to 1964 are about 125 million tonnes of mainly soft itabirite, with at least 50% Fe, and are based on ground-magnetic and gravimetric survey data only.

The high-grade reserve was estimated at about 40 million tonnes including hard laterites and rocks near the surface in three zones, with a total length of about 2500 m, a width of about 100 m, and a depth of about 40 m. The iron ore contents of the corresponding iron oxide- bearing zones in the unweathered basement ranged from 20 to 50%, at an average of about 35%. A drilling program of 27 diamond holes was executed at the time.

Salzgitter Industriebau Gesellschaft studied the feasibility of exploitation in 1966. A look at the current market demonstrates that significant extra tons are required at an improved grade to make exploitation feasible.

Suriname possesses billions of tons of lateritic iron ore resources, which are mainly found in two areas: Bakhuis mountains/Adampada-Kabalebo area (~5 billion tonnes) and near Brokopondo Lake (~1.5–2 billion tonnes). They are generally low-grade in iron (30–35%) and relatively rich in contaminants like Al_2O_3 (25–30% mainly but at least 10–15%) (Doeve 1966; De Vletter 1984; Wong et al. 1998). The primary rocks are banded granulites from the Falawatra Group and volcano-sedimentary rocks from the Marowijne Group, respectively (Wong et al. 1998).

8.2.2 Economic Evaluation

The global market for iron ore is using 62% fines CFR China as standard for pricing reflected in the price chart shown in Fig. 8.1 below. The major producers of iron ore of this quality are BHP, Rio Tinto and Fortescue in the Pilbara region (northwest Australia), and Vale in Para and Minas Gerais States of Brazil. These suppliers represent the first 2 quartiles of the cash cost curve (see Fig. 8.2).

New projects still under investigation include large deposits in Guinea, which would also require large investments in rail and port infrastructure. Those deposits are Simandou and Nimba and have >2 billion tonnes at 65% and >600 million tonnes

Fig. 8.1 10-year iron ore price in US$ per tonne of iron ore fines at 62% CFR China (with January data points from indexmundi.com)

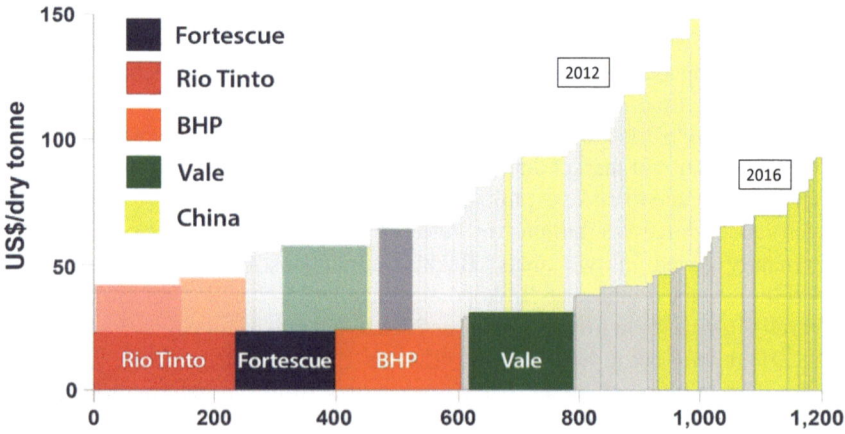

Fig. 8.2 Cash cost curve for iron ore producers 2016 (fortescue.com)

at 64% in iron ore resources respectively, with low alumina and phosphor. These projects are so far not considered economic in terms of acceptable financial risks.

To be competitive in the global iron ore export market, Suriname's iron ore mine would have to be similar in profile to the competitors in the two lowest cost quartiles, which is unlikely, given its location and lack of infrastructure.

The logistics component in operating cost is significant, standing at about 50% of the total delivered cost (McKinsey), hence the more remote mines are disadvantaged even more. Besides entailing high operating costs, it would take large investments in infrastructure to access and to transport ore from south Suriname.

With an estimated rail construction cost of about 6 million US$ per kilometer of heavy duty rail for 25,000 tonne contingents (Chaigneaux, personal communication), the railway line from Tapaje Creek (see Fig. 7.1) to the coast would cover a distance of 380 km. This would entail an investment of: 380 km × 6 M/km = 2.5 billion US$ in rail capital, plus the construction of a deep-sea port at 2.5 billion US$. This port would have to compete with Ponta da Madeira in Brazil, where Vale loads its ultra-large Valemax vessels.

Using Paranam port for export would reduce the maximum capacity vessels to the standard Handymax size of 50,000 dwt and 35 ft draught, and a maximum length of 220 m. With a shorter rail distance of 340 km, the investment in construction of new rail infrastructure and upgrade of existing port infrastructure cuts the investment in half, but it also increases the cost per tonne, due to suboptimal logistics.

Logomerac concluded, in 1965, that export of iron ore was not a viable option, but he proposed the smelting of laterites in low-shaft electric furnaces, combining the presence of large laterite resources with hydro-power and charcoal. In a subsequent report, he suggested a similar process for the treatment of the residues, stemming from decades of alumina refining stored at Paranam. In his view, a range of valuable products could be produced, from titanium and alumina to REEs and radio-active elements (De Vletter 1984).

Professor Velzeboer of TU Delft suggested to De Vletter, in 1982, to locally produce high-value ferro-alloys, by using local manganese and chromite as in-put for the process.

The laterites at Bakhuis/Adampada/Kabalebo (as at Brokopondo) contain iron minerals in grain size of a few microns, intergrown with other constituents, making concentration complicated (Bosma et al. 1973). If efficient treatment of these ores would be possible economically, and if cheap power would be available, the combined output of alumina and iron concentrate from the Bakhuis area would be significant. The Kabalebo hydroelectric power project was intended to provide cheap energy to accommodate processing, but its realisation was very doubtful in Wong et al. (1998), and still is today. The recent oil discoveries in continental waters could provide the missing link.

A combined business case of mining and processing lateritic iron ore, bauxite (described in Chap. 6), lateritic nickel (in Sect. 7.3) and phosphate (in Sect. 9.6) could justify a large capital investment in infrastructure and power. Alternatively, lateritic iron ore and nickel with phosphate could become economic if Bakhuis is developed for bauxite, after all.

8.3 Manganese

Every tonne of normal steel requires about 10 kg of manganese, because it assists in de-oxidation of steel and prevents formation of iron sulfide. Manganese steel -or mangalloy- is a popular alloy containing 11–14% manganese, and is known for its high impact strength and extreme anti-wear properties. Besides possessing healthy demand in the above-mentioned steel applications, manganese will also enjoy a significant additional up-side, coming from clean-energy applications, since manganese is widely used as a battery component.

Manganese is found in all rocks such as granites, schists and sandstones, though in minute amounts. The number of manganiferous minerals occurring in nature is more than 150, but there are not many wide-spread manganese-rich minerals. The chief manganese minerals of commercial importance are oxides, although with the exhaustion of oxide ores, carbonate ores are becoming increasingly important.

Between 1954 and 1966, deposits and indications of manganese were discovered and explored. They were associated with metamorphic sediments, like gondites of the Paramaka formation, as discussed in Sect. 8.2. These gondites are important for secondary enrichment, but have no economic importance on their own. Manganese deposits in Suriname were the subject of a comprehensive Ph.D. research study by Prof. Holtrop, between 1959 and 1962.

He stated that, in general, the following factors are of importance for the enrichment of manganese ore:

- A large quantity of proto-ore, defined as the mother rock from which manganese compounds are dissolved.
- Rhodochrosite, manganese carbonate and a favourable proto ore which can dissolve and precipitate as manganese oxide, as at the Serra do Navia deposit in Brazil.
- Many thin layers of gondites for a large erosion contact surface.
- Vertical position of the proto orebody, to enable deeper ground-water penetration and oxidation.

8.3.1 Exploration Results

Deposits of some significance identified in Suriname are listed below, and are shown in Fig. 8.3.

- Maripa Hill and Pique Hill on the eastern shore of the Brokopondo Lake
- Apoema Soela on the west bank of Marowijne river
- Lada Soela south west of Paloemeu
- Plet Ridge east of the Emma Mountain range.

Maripa Hill was first discovered by a gold explorer in 1954, but could not be traced back until Doeve rediscovered this hill in 1956, with its outcrop of layered manganiferous material, which appeared to be highly weathered gondite (Wong et al. 1998). It has been studied by Holtrop in much detail, as part of above-mentioned study. Selective weathering on the hill persists to depths of about 80 m, shown by 9 diamond holes, with the best part on top of the hill, where residual ore is found over a distance of 2 km (Bosma et al. 1973).

Holtrop estimated the following reserves: (a) 726 kt of unweathered and slightly weathered proto-ore (spessartite rock), with an average Mn content of 24.3%, embedded in 7.3 Mt of clay, and (b) 530 kt at 28.2% Mn of fragmental residual ore, including nodular ore, mainly present in the upper 3 m.

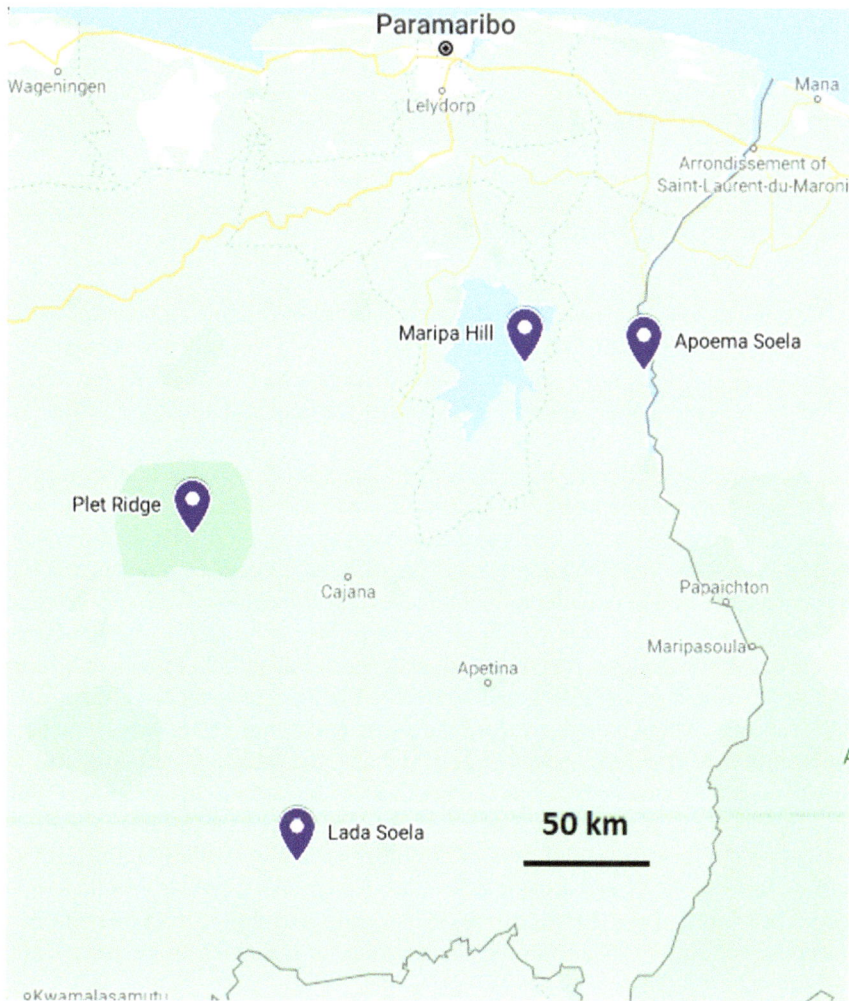

Fig. 8.3 Location of known manganese deposits

It was identified as the largest deposit with the highest economic potential. Its location is quite favourable, standing at a distance of about 40 km from Nassau and 50 km from Afobakka, thus with access to water and power. It requires significantly more resources at similar grades to become feasible, as will be demonstrated further. Figure 8.4 shows a profile of the deposit as assumed by Prof. Holtrop.

Several small lateritic manganese deposits of apparent ore grade were discovered on a ridge of about 6 km in length, running from Apoema Soela at the Marowijne River to the west. They occur in the same metavolcanic-sedimentary unit as Maripa Hill (Holtrop 1962 and De Vletter 1984).

Maripa Heuvel

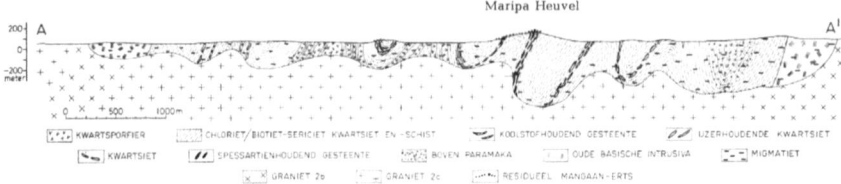

Fig. 8.4 West to East profile of Maripa Hill deposit

According to a rough estimate, the total quantity of residual and detrital mantle ore does not exceed 300–500 kt, and is probably less. Generally Mn contents vary between 35 and 45%, but contamination with fragmentary iron laterite reduces the average Mn content of the material recovered by washing to about 30% (Bosma et al. 1973).

The Lada Soela deposit, near the Upper Tapanahony River, was discovered in 1966, after Salzgitter Industriebau Gesellschaft had carried out its study on possible sources of iron ore for the Government of Suriname in 1966. The deposit, which occurs in an 8 km-long zone of metamorphic rocks surrounded by granites, has been systematically explored, with pits and trenches and 108 diamond drill holes through 80 m thick weathered zones on hill tops. The resources of lateritic manganese ore in the enriched, southern zone has been estimated at about 500 kt with an average (total) Mn content of about 40% and an average Mn/Fe ratio of about 5. This consists of (a) about 330 kt of mantle ore, forming a laterite cover of 250 m length and 6.9 m thick at 41% Mn, and (b) about 160 kt of soil near the surface at 38% Mn. This top layer is contaminated with blocks and nodules of iron laterite (Bosma et al. 1973).

The Plet Ridge deposit was discovered in 1963, and is located east of the Emma range near the Toekoemoetoe Creek, off the Saramacca River, in the Central Suriname Nature Reserve. It has an extension of less than 0.5 km, and is connected with a varied series of mainly metasedimentary rocks, including spessartite, quartzites and basic metamorphic rocks. They form the southern part of a xenolith in granite of about 2.5 km in length (Bisschops 1969).

Based on 12 diamond holes and 47 auger drill holes, the deposit is estimated to contain 150 kt, with an average Mn content of 35.4% and a Mn/Fe ratio of 8:4. The size of the resource, in combination with the inaccessibility of the Toekoemoetoe area and the environmental restrictions, make this resource unfeasible for exploitation, considering the above parameters. However, this deposit is rich in manganese carbonates, and is currently subject to further research. Economics may change, if sufficiently large quantities of carbonates of commercial grade (see below) can be added to the resource.

8.3.2 Economic Evaluation

For a perspective on economic potential, it makes sense to compare size and grades of resources against the world's suppliers as shown in Table 8.1.

Serro de Navio produced 31 million tonnes of washed oxide ore, assaying from 42 to 49% Mn, 1.8 million tonnes of manganese pellets, sinter feed and fines, and 0.9 million tonnes of carbonate ore, assaying about 30% Mn during its lifetime. The mine was located in Amapa state in Brazil on the Guiana Shield, and produced manganese ore from 1959 to 1998. For the first 30 years, this was one of the world's largest manganese ore mines, but output declined substantially since the mid-1980s (Drummond 1999; Scarpelli 2018).

For reference, two manganese mines from the above table will be discussed in more detail. Nsuta in Ghana and Matthews Ridge in Guyana are selected based on their relevance for Suriname: both mines are situated in locations geologically related to Suriname.

Table 8.1 World manganese producers. *Source* Callabonna Resources, 2014

Name	Location	Mn Ore (Mt)[b]	Grade (% Mn)	Method	Comment
Tambao	Burkina Faso	36	48	OP	
Groote Eylandt	Australia	75.6	46	OP	
Wessels[a]	South Africa	64	44	UG	
Moanda	Gabon	100	43	OP	
Nchwaning/Gloria[a]	South Africa	165	42	UG	
Gravenhage[a]	South Africa	13	40	UG/OP	
Woodie Woodie	Australia	16.7	38	OP	
Mamatwan[a]	South Africa	46	37	OP	
Tshipi[a]	South Africa	60.5	37	OP	
Azul	Brazil	15	35	OP	
Nsuta	Ghana	24.4[c]	29	OP	
Zhairem	Kazakhstan	54.2	24	OP	
Serra do Navio	Brazil		49	OP	Mined out
Matthews Ridge	Gyuana	29.2	14	OP	In construction

[a]These mines are all located in the Kalahari Manganese Basin, an area of 400 km^2 containing 13 billion tonnes of manganese resources at 20–48% Mn
[b]Reported reserves
[c]As per 2011 for more details see text below

8.3.3 Nsuta Manganese Deposit

As stated in Sect. 5.1, the Man-Leo Shield in West Africa hosts several manganese deposits associated with the Birimian series, some of which have been mined, specifically at Nsuta in Ghana (Kessie 1985). Nsuta has been the main source of high-grade manganese ores for over a century. The mine has produced between 55 and 60 million tonnes of ore in its life-time (USGS.com and Kessie 1985), with production averaging 1.5 Mtpa since 2000. Production of oxides has fallen over time, and the high grade ores have been depleted, but the mine is still producing, moving from oxides to carbonates with grades of about 29% Mn, which is a trend of interest for Suriname, specifically for the Plet Ridge deposit.

The mine is connected to Takoradi Port by a 60 km long railway line. Nsuta is located only 6 km south of Tarkwa, the base of one of Ghana's oldest and largest gold mines. The manganese deposits are metamorphosed sediments which occur on five hills, over a stretch of 4 km in lenticular bodies, up to 300 m in length and 30 m thick, varying greatly in size.

The manganese minerals are chemically of two main types: oxides and carbonates. Pyrolusite and psilomelane are the main oxide minerals, whereas rhodochrosite and dialogite are the main carbonate type minerals underlying the oxide orebody, with a sudden change in the ore layers between the two types.

The most valuable ore is the high-grade oxide, or so-called battery grade ore, because it is used in the manufacture of dry cell batteries; it contains about 52% Mn. The lower grade oxide ore is known as standard metallurgical grade, with a manganese content of 48–50%, used for steel production and lower grades of 46% or B-grade and 42–45% or C-grade. The carbonate ores are not as useful as the oxide ores, but can be valuable when they are of good grade and when phosphorous content is low. High-grade carbonates typically contain about 25% Mn, with low grade between 15 and 20% Mn.

After mining high-grade manganese oxide ores for over 70 years Nsuta's owners started to move towards mining carbonates by the end of the past century. The first step was to calcine manganese carbonates in a rotary kiln to produce marketable manganese oxide nodules. By heating $MnCO_3$, MnO with a grade of about 42% and CO_2 are produced.

The mine is now owned by Consolidated Minerals and is only producing carbonate ore. The latest available resource and reserve information (JORC compliant) is listed below as reported per 2014 (Table 8.2).

Table 8.2 Nsuta reported reserves and resources (www.consmin.com)	Category	Mt	Mn (%)
	Total reserves	45.01	28.16
	Total resources	101.3	26.80

8.3.4 Matthews Ridge

Matthews Ridge can be considered as being on the margins of a viable economic project, as the decision to invest has been reconsidered over a period of about 50 years.

Initial exploration on Matthews Ridge was done in the 1950s, and included pitting and some drilling, as well as extensive geological mapping. The available historical information indicates that Union Carbide operated the Matthews Ridge mine, and shipped concentrates out of Guyana between 1962 and 1968. The records indicate that during that period, some 1.7 million tonnes of manganese concentrate (37% Mn) was recovered and shipped. By the termination of mining activities in 1968, the known resources base was reported to be approximately 1 million tonnes of recoverable concentrate at 37% Mn. At Pipiani, which is south of Matthews Ridge, the historically-calculated resources of available concentrates, based on the mining and beneficiation methods used at Matthews Ridge, were 642 kt of 42% Mn on the basis of 38 drill holes, numerous pits and trenches. The total resource was further increased to 3.6 million tonnes of 33.4% Mn recoverable concentrate, based on exploration conducted by the Guyana Geology and Mines Commission and North Korea in 1985. Reunion Manganese Inc., a subsidiary of Reunion Gold at the time, carried out an exploration program to identify and measure the manganese resources. The work was initiated along an area 15 km in length, encompassing nine hills, partially mined in the 1960s.

The exploration program for the mineralised saprolite of this area begun in 2010, with the mechanised trenching of all hills. The trenching program was completed in 2011, with 130 trenches. A total of 735 diamond drill holes and 235 reverse circulation holes were drilled (Reunion, 2013).

At a cut-off grade of 8% Mn, the unconstrained measured mineral resource was estimated at 17 Mt at 14% Mn. The non-diluted indicated mineral resource was estimated at 15 Mt, averaging 13% Mn. The unconstrained inferred mineral resource was estimated at 3 Mt at 15% Mn.

The vertical profile typical of Matthews Ridge shows three manganese-bearing units covered by a 100 m-thick, sterile upper phyllitic unit with a sharp contact. The enrichment occurs thanks to a mixture of circumstance with large quantities of gondites and strong chemical weathering. Braunite occurred as proto ore, which formed the high-grade manganese oxide (Holtrop 1962).

In 2016, Reunion Gold Corporation sold all of its rights in the Matthews Ridge manganese project to Bosai Minerals Group Co, a Chinese corporation. The new owner has set a time-frame of July 2019, to develop the mine and processing plant in order to begin extraction, and is investing approximately US$100M in this stage of development.

The Matthews Ridge case demonstrates the fact that the known resources in Suriname are insufficient for economic exploitation. Manganese prices have been fairly stable, in recent years, as shown in Fig. 8.5, but manganese can potentially be a high-margin business, especially in a boom, with prices more than double, as they were in 2008. In general, the world has gone short of high-grade manganese from

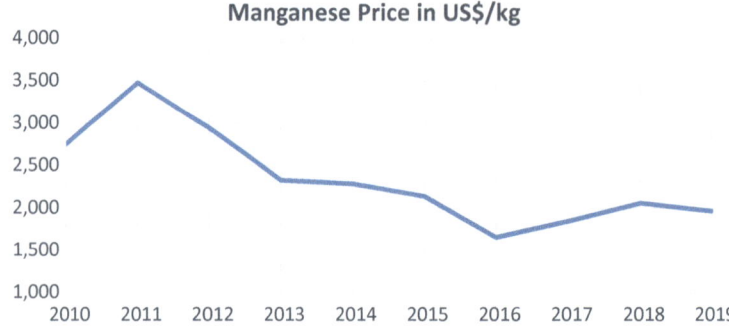

Fig. 8.5 10-year manganese prices (with January data points from Metalary.com)

time to time, so it is worth investigating if extensions and/or carbonates are present at greater depth, specifically in accessible locations, which might perhaps match or exceed the resources at Matthews Ridge.

8.4   Molybdenum

Molybdenum is primarily used as an alloying element in steel and superalloys, where it adds hardness, toughness and resistance to corrosion and abrasion, in combination with adding strength at high temperatures. Applications are resident in cutting tools, and for equipment in the production of oil and gas, thanks to its resistance to seawater and to severe chemical environments.

A significant fraction of molybdenum comes as by-product from porphyry copper mining. Nearly all of the known molybdenum ores consist of molybdenite or MoS_2 (Craig-David et al. 1996).

8.4.1   Exploration Results

Molybdenite has been found in cores drilled near Dramhoso, and other occurrences have also been reported from veins of granite southeast of the Plet Ridge manganese deposit (Bosma et al. 1973).

8.4.2   Economic Evaluation

Given the fact that molybdenum is produced in large copper mines as a by-product, albeit at modest extra cost, this makes it difficult to enter the market successfully unless exceptional grades of the ore are found. For an illustration of this, Bingham Canyon mine in Utah produces 30 million lb of molybdenum per year, as a by-product from its copper-gold-silver operation.

The price averages 30 US$/kg for 2019. Molybdenum reached an all-time high of 87, in September of 2005, and a record low of 12.40 in November of 2015.

References

Bisschops JH (1969) An occurrence of chromite near Emma Range in Central Suriname. In: Proceedings 7th Guiana Geological Conference, Verhandelingen Koninklijk Nederlands Geologisch Mijnbouwkundig Genootschap, vol 27, pp 119–124

Bosma W, Ho Len Fat AG, Welter CC (1973) Minerals and mining in suriname. Mededelingen Geologisch Mijnbouwkundige Dienst Suriname 22:71–101

Craig-David JR, Vaughan DJ, Skinner BJ (1996) Resources of the earth: origin, use and environmental impact. Prentice Hall, p 472

De Vletter DR (1984) Economic geology and mineral potential of Suriname. Mededelingen Geologisch Mijnbouwkundige Dienst Suriname 27:11–30

Den Hengst P (1975) The Upper Saramacca chromite. Mededelingen Geologisch Mijnbouwkundige Dienst Suriname 23:244–249

Doeve G (1966) Delfstoffen in Suriname. Mededeling 15 van Geologisch Mijnbouwkundige dienst van Suriname, 94–107

Drummond J (1999) Private investments, environmental impacts and quality of life in a tropical mining venture: the case of the Serra do Navio manganese mine (Amapá, Brazil). In: Conference of the American society for environmental history, Tucson, Arizona, US

Fortescue 2016 Annual General Meeting presentation, slide 17–www.fmgl.com.au

Holtrop JW (1962) De mangaanafzettingen van het Guyana schild, Thesis Delft. Also published (1962) in Mededelingen Geologisch Mijnbouwkundige Dienst Suriname 13:514 p

Kessie GO (1985) The mineral and rock resources of Ghana–published by A.A. Balkema, Rotterdam/Boston, p 610

Ornstein MAM, Haug GMW (1965) A recent discovery of itabirite in Suriname. Geologie en Mijnbouw 42:359–363

Reunion Gold Corporation (2013) NI43-101 Technical report on Matthews Ridge Manganese project

Scarpelli W (2018) Chromium, iron, gold and manganese in Amapá and northern Pará, Brazil. Braz J Geol 48(3). São Paulo July/Sept 2018

Wong ThE, de Vletter DR, Krook L, Zonneveld JIS, van Loon AJ (eds) (1998) The history of earth sciences in Suriname, 479 pp

Chapter 9
Critical Raw Materials

Since 2011, the European Commission has created and updated a list of Critical Raw Materials (CRM). The list contains highly sought-after metals and minerals, thanks to accelerating technological innovation cycles and to the rapid growth of emerging economies. These CRMs are particularly important for high tech products and emerging innovations, designed to improve the quality of life. They are irreplaceable in the manufacturing of smartphones, solar panels, wind turbines, electric vehicles and energy-efficient lighting, and are therefore considered to be very relevant for improving the environment. Driven by targets to achieve energy transition goals, demand for certain raw materials is expected to increase by a factor of 20 by 2030. In June 2017, the World Bank released the report The Growing Role Minerals and Metals for a Low Carbon Future and concluded that such a future would be very mineral-intensive (World Bank report 2017).

The EU list of Critical Raw Materials contains raw materials which reach or exceed thresholds for both economic importance and for supply risk (https://ec.europa.eu). The latest update took place in 2017, and resulted in the following list:

The metals in bold have been identified in Suriname, and will be discussed in more detail in the following chapter (Table 9.1).

9.1 Beryllium

Beryllium (Be) is one of the lightest of all metals, and has one of the highest melting points of any light metal. Its main use is for beryllium–copper alloys in electronics. Thanks to its extreme lightness, it is used in aerospace, defense and science applications. Its neutron- reflecting properties are utilised in nuclear reactor control rods.

Beryl ($Be_3Al_2(Si_6O_{18})$), a silicate mineral containing 12–13.5% beryllium oxide (BeO), is one of the main sources of beryllium. World mine production in 2018 was about 230 tonne (USGS, 2019), mainly from the United States of America, and with contributions less than 10 t from several countries, including 3 t from Brazil. A further

Table 9.1 2017 list of CRM as per EU Commission

Critical raw materials			
Antimony	Fluorspar	**LREEs**	Phosphorus
Barite	Gallium	Magnesium	Scandium
Beryllium	Germanium	Natural graphite	Silicon metal
Bismuth	Hafnium	Natural rubber	**Tantalum**
Borate	Helium	**Niobium**	**Tungsten**
Cobalt	**HREEs**	**PGMs**	Vanadium
Coking coal	Indium	**Phosphate rock**	

20–25% of consumption occurs from recycling. Gem varieties include aquamarine and emerald.

Near Rama and Phedra, on the banks of the Suriname River, pegmatites containing beryl have been found on both sides of the river. Mining took place from a 25 m wide outcrop of the eastern pegmatite. Between 1953 and 1955, 9 t of beryl was mined, with a BeO content of 11.7% (Bosma et al. 1973; Wong et al. 1998). Production took place in the part of the pegmatite above water level, where 3,000 m^3 of weathered pegmatite was monitored and washed, corresponding with a grade of 0.13%. Mining has since been abandoned, due to insufficient grade, but the pegmatite persists, both under the level of the river and on the western side of the river, with estimated grades of 0.04–0.06%. Pegmatites are discussed in more detail in Paragraph 8.6.

Although these numbers may seem minute, one should note that beryllium metal is currently trading at 830 US$/kg on the Shanghai Metal Exchange, which can make a small scale operation feasible.

9.2   Cobalt

Cobalt is an important component in super-alloys that retain their mechanical strength at high temperature, and are resistant to the corrosion of hot gases. Super-alloys are used in jet engines and gas turbines. Cobalt alloys also have very strong magnetic properties.

Demand for cobalt is strongly increasing, thanks to energy transition initiatives driving demand for electric vehicles (see also Paragraph 7.2 on copper). Demand for batteries makes up approximately 40% of the total cobalt market. Total demand will grow by almost 10% per annum, reaching 210 kt by 2025 (Mining Magazine, June 2018). Global cobalt production needs to more than double from 94 kt refined cobalt in 2016 in order to reach 210 kt by 2025. Bloomberg predicts a steady increase in cobalt demand to around 400 kt by 2040 (Ong 2017).

Democratic Republic of Congo (DRC) is supplying 58% of world production, as per 2017, and possesses rich resources, which will make it a key supplier in the foreseeable future, despite reports of human rights abuses. However, the country is politically unstable, and therefore investors in its mining sector are cautious. Other suppliers may develop, while the users of cobalt may look for substitutions.

9.2.1   Exploration Results

In the period of 1952–1953, Billiton Maatschappij Suriname undertook exploration for asbolane in an area east of the Suriname River, between Brokopondo island and Balling Soela. Asbolane is a poorly-defined mineral, also referred to as cobalt-bearing "wad" (mindat.org). It is a mixed-layer mineral, with layers of MnO octahedra and of other metals, mostly occurring in separate layers. During the exploration program, 6 tonnes of asbolane-sample was collected and analysed, with an average grade of 188 g Co per m^3.

Minor amounts of cobalt were associated with Adampada lateritic nickel, as discussed in Paragraph 7.3 (De Vletter 1984; Wong et al. 1998).

9.2.2   Economic Evaluation

At the average five year price of 25 $/lb (see Fig. 9.1), 1 m^3 would generate 10 US$. At current prices, large volumes are required (or much higher grades) to make a case for economic extraction.

9.3   Heavy and Light Rare-Earth Elements

The Rare-Earth Elements or REEs consist of the lanthanide elements in the periodic system, plus the chemically similar elements scandium and yttrium. These elements

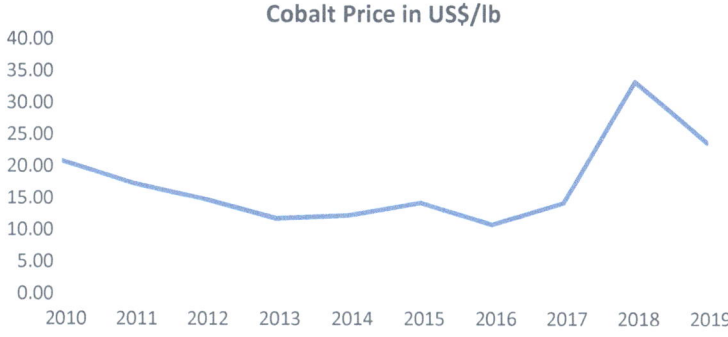

Fig. 9.1 Ten-year cobalt price (with January data points from https://www.investmentmine.com)

are actually not specifically rare (in its literal sense) and are described in detail in The Rare Earth Elements: An Introduction by J. Voncken, in the Springer Briefs in Earth Science series (Voncken 2016).

The light REEs or LREEs are the group of elements ranging from Cerium to Gadolinium, and the heavy REEs or HREEs are the group located to their right, ranging from Terbium to Lutetium, plus the 2 above-mentioned elements.

Demand is driven by the growing interest in renewable and more efficient energy technologies, which rely, among others, on the magnetic and fluorescent properties of HREEs. They have taken a more prominent role than the LREEs in recent years, but are generally less abundant in REE deposits (Barakos 2018). It is not possible to selectively mine a specific element; they have to be mined together, and also must be separated through metallurgical processes. China has taken the lead in developing technologically advanced separation methods, which offers them a competitive advantage in processing REEs.

9.3.1 Exploration Results

Research of anomalies identified during airborne magnetic and electromagnetic surveys in the Upper Nickerie River area of the Bakhuis mountains indicated the presence of REE associated with apatites (group of phosphate minerals), nickel and copper. It was advised to undertake additional geochemical surveying of soil and bedrock and drilling at specified target areas, in view of the expected presence of carbonatite at depth (Dahlberg 1982).

Paragraph 9.5 will discuss the potential of phosphate, which is noteworthy in this regard.

9.3.2 Economic Evaluation

Annual global consumption in 2017 stood at 150 kt, which shows that the market is rather small, and can be sensitive to oversupply and price manipulations. In terms of supply, China is (and has been) dominating the market for years. China has managed to reduce the prices to a level low enough to prevent other potential producers from entering the market, but there are signs that their potential to influence prices is diminishing.

Projects that can start with a low capital investment and then run at low cost may have a distinct advantage, as well as mines which produce REE as by-product. Recent research in the United States points out that the size and concentration of HREEs as by-products in some unmined phosphorites predominate over the world's richest REE deposits (Barakos et al. 2016).

9.4 Platinum Group Metals (PGM)

The six platinum group metals—platinum, palladium, rhodium, iridium, ruthenium and osmium—always occur together in the same geological setting. Many of their chemical and physical properties are similar. Platinum is the most abundant member, hence the name. Platinum metal occurs in its native form in placers. All of the PGM are resistant to corrosion; each has a high melting temperature, and each has interesting properties as a catalyst. Their efficacy in speeding chemical reactions, plus their use in highly corrosive environments and very high-temperature situations, such as in automobiles, account for the main uses of all PGM (Craig-David et al. 1996).

World PGM reserves total about 3200 M oz (~200 years at the current rate of production), and are found in relatively few areas of the world: South Africa (88%), Russia (8%), North America (2%), Zimbabwe (1%) and Rest of World (1%).

South Africa (SA) is the source of over 60% of newly mined PGMs and over 80% of Platinum.

9.4.1 Exploration Results

First in 1938, and later by miners (and during prospecting expeditions by GMD), alluvial platina was only found together with gold in various creeks south of De Goeje mountain range, specifically in Platinum Creek, Telemac creek and Sand Creek, with Latitude-Longitude coordinates 3°20′ and 54°10′ (GMD 1956–58). Based on gravel samples, it was then estimated that the content of platinum roughly amounted to one-third of that of gold (Bosma et al. 1973).

The primary source has not (as yet) been identified, but the Platinum Creek runs through a valley with fairly steep slopes 25 m high and with quartz outcrops. Geologists suspected this to be the source of most of the platina recovered in gravels (Van Kallen. 1956).

9.4.2 Economic Evaluation

The price for platinum has been on a steady decline since 2011 as shown in Fig. 9.2 below.

The similarities are obvious, so for the economic assessment of a PGM deposit in Suriname the author refers to Chap. 5 on Gold.

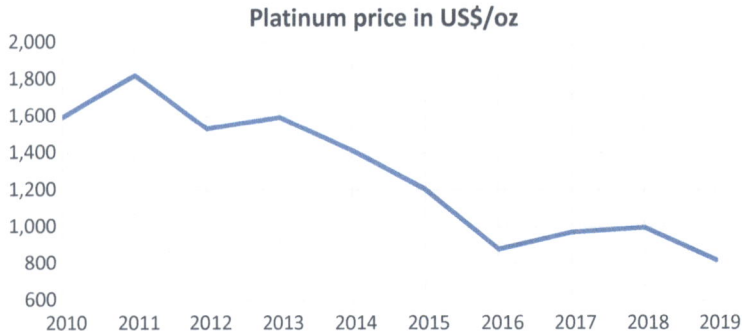

Fig. 9.2   Ten-year prices of platinum (with Janury data points from https://www.Macrotrends.com)

9.5   Phosphate

Phosphate P_2O_5 is the oxide form of phosphorous. The earth's crust contains about 0.23% phosphate, mostly appearing as apatite. Bone is another occurrence of phosphate. In the 1600s, English farmers observed that ground bone increased their crop yield. In later centuries, European countries imported bones from every available source, until, in the 19th century, techniques were discovered to convert natural phosphate minerals into usable fertilizer forms (Craig-David et al. 1996). There is no substitute for phosphate fertilizers, a fact which explains its strategic value in feeding a growing world population. The major world suppliers of phosphate are USA, Morocco, China and Russia.

9.5.1   Exploration Results

Exploration programs on the Bakhuis Mountain range carried out by GMD resulted in the discovery of apatite in lenses in the banded gneisses. Apatite is the world's major source of phosphorus, which is one of the main nutrients in fertilizer.

Phosphate mineralization was observed in diamond drill cores. A total of 63 holes, averaging 50 m in depth, were drilled in the period between 1975–1982, to follow up on a base-metal exploration program in the Lower Proterozoic, high-grade metamorphic Bakhuis zone, where previous airborne magnetic and electromagnetic investigations have established several anomalies. Drill cores with an estimated content of more than 3 volume-% of apatite were analysed over stretches of two metres; the analyses give P_2O_5 values up to 14.4%. The apatite occurrences coincide partly with copper mineralization, but the highest P2O5 contents were found outside the zone, with copper mineralization (Dahlberg 1987).

These apatite lenses probably are metamorphosed sedimentary phosphorites, and could potentially make up a widespread body of considerable size and extent. The widespread distribution of banded gneisses granulites, which contain the apatite in Area XVI, about 15 km SE from the railhead at Bakhuis, give evidence that the apatite is widespread (Sheldon 1982) (see also Fig. 7.2).

It should be noted that the areas in Roman figures differ from those in arabic, as used in earlier reports on Bakhuis' bauxite. Area XVI is located in bauxite area 21 in the southern half of the bauxite resource.

The above deposit is also referred to as K/3 deposit, named after a helicopter pad near the Nickerie River; the deposit is located at latitude 4° 26′ N and longitude 57° 04′ W. The length of the anomaly found is about 8 km, and is open at both ends, with a probable width of the apatite zone of 50 m and thickness of 2–3 m (Dahlberg 2005). The potential for extension over a large area is supported by a small amount of apatite, found in a core drilled near Three Falls at about 20 km distance (Sheldon 1982).

The crystalline and chemical nature of the apatite appears favorable for beneficiation into high-quality apatite concentrate. Based on his assessment that the potential of the deposit was great enough to proceed with further studies, Sheldon proposed a program to further delineate the resources of apatite-rich rock in a wider area. If zircon grains present in existing thin sections are rounded, then the rock is of metasedimentary origin, in which case it can be expected that the apatite occurs widespread in the Bakhuis Mountains (Sheldon 1982). Preliminary work by the author indicates that this scenario could be likely, following analysis of zircons present in thin sections (made available by De Roever).

9.5.2 Economic Evaluation

If these apatite resources would prove adequate, there is potential to develop a mine and/or concentration process. The products of an apatite mining and processing industry could be:

- Apatite concentrate (phosphate rock) for export and domestic use for elemental phosphorus manufacturing
- High-grade elemental phosphorus for a chemical and fertilizer industry
- Fertilizer manufacturing for domestic use and export.

Production of phosphorus from apatite requires an electric furnace, thus availability of cheap power would provide a competitive advantage. If the infrastructure at Bakhuis would be re-developed for bauxite mining and processing, including the development of a coastal refinery, then the production of apatite and phosphorus could become economic, and bring additional returns.

Morocco and Western Sahara are important producers of phosphate rock, at an estimated 33 million tonnes (as reported in 2018). For pricing, Morocco ore serves as reference, averaging 104 US$/t FOB North-Africa for 30% P_2O_5 over the last five years.

9.6   Tantalum/Niobium (Columbium)/Lithium/Tin

Columbite and tantalite, the minerals containing niobium (or columbium – see below) and tantalum occur in pegmatites and in alluvial gravels which originate from them. This group of metals occurs locally in pegmatite minerals, in combination with cassiterite (a source for tin) and amblygonite (formula $(Li,Na)AlPO_4(F,OH)$ is Suriname's only known source for lithium), and are therefore described in one paragraph.

Two types of pegmatites are recognised:

- Zoned pegmatites with or without minerals of the columbite/cassiterite series
- Non-zoned, mostly sterile pegmatites.

The zoned pegmatites have a cylindrical shape, containing albite and minerals with potential value. Sterile pegmatites are known from different areas in the country, but the ore-bearing pegmatites found to date are concentrated in a 20 km wide zone, from Kwakoegron towards Bigiston, including Phedra on the Suriname River and the Jorka Creek area near Marowijne River.

Tantalum is named after the Greek king Tantalus, a mythic figure who was forced to stand in knee deep water which he could not drink, with delicious fruit hanging overhead and just out of reach—all by way of tortuous punishment. The name refers to tantalum's ability to be submerged in substances without being dissolved. It makes extremely hard alloys, for use in cutting tools.

Tantalum's unusual characteristics led to its increased use in the late 20th and early 21st century. The element is extremely stable at temperatures lower than 150 degrees Celsius, and needs exposure to hydrofluoric acid to cause corrosion. This protection from corrosion is due to a natural protective layer, created by oxides of tantalum on the surface of the metal. Tantalum is also used to create surface acoustic wave filters, devices used in cell phones and televisions to improve audio quality. The average cell phone contains about 40 milligrams of tantalum.

Niobium was referred to as columbium in U.S. literature, from the time of its discovery in 1801 until about 1950, when the world's chemists agreed upon the designation "niobium." The USGS and some metallurgists still reference the historic name, and any material published prior to the early 1950 s will use the term "columbium."

As a pure metal, niobium is ductile, white, shiny and soft. It has a number of useful qualities.

In particular:

- It is superconductive
- It gives resistance to corrosion, strength and toughness, when used in alloys.

In the early days after its discovery, niobium was often indistinguishable from tantalum, and they were somewhat like identical twins; hence its name as an offspring of Tantalus. In Greek mythology, Niobe is the daughter of Tantalus.

Niobium and tantalum often occur together in nature, and are mined as coltan by small-scale operators, partly in conflict zones in Africa, a fact which leads to consumers demanding "conflict-free" sources.

The entire world's supply of mined niobium comes from three operating mines: two in Brazil and one in Canada. The largest and highest-grade operation, Araxá in Minas Gerais, Brazil, dominates the market, with ~85% of global production (Ronald 2016).

Lithium nowadays enjoys fast growing demand, thanks to its use in batteries in electric vehicles, at an expected rate of 14.6% per annum going up to 2025 from 185 kt lithium carbonate, equivalent to about 632 kt (McKinsey).

Lithium carbonate production shifted from pegmatites to brines (from salas and playas in South-America and USA), closing non-zoned pegmatite operations in the process. Certain applications remain for lithium in the form of oxides from zoned pegmatites in the ceramic and glass industries.

Tin is mainly used in the solder industry; around 55% of the supply is used for this purpose. It is also widely used as plating for steel, especially for food preservation.

9.6.1 Exploration Results

In 1956, columbite/tantalite was found in alluvial material of the small tributaries of the Jorka Creek and the Granite Creek—a tributary of the Toemoffo Creek. The grade of samples ranged from traces to 500 g/m^3 of gravel, with gravel layers of decimeters and overburden up to 1 m. In the Tempati Creek and Patamacca Creek areas, columbite and tantalite were found with the highest grade about 200 g/m^3 (Van Eijk 1961). The annual report (1956–1958) of the GMD also mentions indications of columbite and tantalite from pegmatites west of Phedra.

Figures 9.3 and 9.4 show samples of columbite and tantalite in the Marwijne area.

In 1961, Billiton Maatschappij Suriname started exploration in a small area close to the Marowijne River, a few kilometers southwest of the confluence with Jorka Creek, about 35 km upstream from Albina, also referred to as EP4. Here, the outcrop of a small Li-Ta-Sn pegmatite body was observed in heavily weathered pre-Cambrian rocks, during a field visit in 1956 (van Eijk, Yearbook 1956–1958). As part of its exploration program, Billiton carried out a pilot exploitation of amblygonite and tantalite/cassiterite contained in the near surface part of this body. Amblygonite appeared in enormous masses of up to 8 m. A sample is shown in Fig. 9.5. Heavy chemical weathering hampered the study of the internal structure.

A total of 8,500 m^3 of weathered pegmatite and host rock were processed at an equivalent weight of about 21,000 t. The total quantity of washed amblygonite was 1700 t at 8% Li_2O. The total quantity of combined tantalite/cassiterite concentrate was 20 t. This operation was terminated in 1962, after the conclusion was reached that the deposit was too small for economic exploitation; numerous small pegmatite bodies were found, but without any trace of lithium, and only very low percentages of tantalite and cassiterite (Montagne 1964).

Fig. 9.3 Columbite sample from Albina (Naturalis Biodiversity Center, Leiden)

Fig. 9.4 Tantalite from Marowijne area (RGM.1070004; Naturalis Biodiversity Center, Leiden)

Fig. 9.5 Sample of amblygonite from Jorka Creek (Naturalis Biodiversity Center, Leiden)

In 1970 and 1971, eluvial material containing 100–1,000 g/m^3 of combined ore was domestically mined and processed, and about 2.4 tonne of a mix containing 40% cassiterite and 40% tantalite was exported.

Beside the above occurrences, cassiterite appeared in the same Jorka Creek area, in a zoned pegmatite named HL 8, as well as in a number of pegmatites in the Patamacca Creek area.

9.6.2 Economic Evaluation

Opportunities for large-scale exploitation are not foreseen. Operations like Mt Cattlin in Western Australia, which re-opened recently, process 800,000 tonne per year, requiring an investment over 100 million US$ in access, plant and equipment, and a reserve of 10 Mt. Mt Cattlin reports these tonnages at 1% LiO$_2$ and 150 g/t Ta$_2$O$_5$.

Thanks to strong demand in the high-tech industry and high commodity prices, development of a smaller-scale mine could be feasible.

The current prices average (as per Metal bulletin and Infomine, 2019):

- 75$/lb for Ta$_2O_5$
- 40–45$/kg for niobium
- 700$/t of LiO$_2$ (5–6%)
- 10$/lb for tin.

Processing small volumes with a simple gravity separation plant can work well, thanks to the high specific density involved. A site visit by the author confirmed accessibility via the Marowijne river adequate for small-medium scale development, without the need for costly road construction to Albina.

A 1,000 tonne per day operation has the potential to generate over 10 million US$ per year in revenue, if sufficient tonnages at historic grades can be found. Thus, exploring extensions of the pegmatites downward could potentially add sufficient reserves for small-scale success.

9.7   Tungsten

Tungsten, or wolfram, is a chemical element with symbol W. The name tungsten comes from the former Swedish name for the tungstate mineral scheelite, tung sten or "heavy stone." Tungsten is a rare metal found naturally on Earth, and it is almost exclusively combined with other elements in chemical compounds rather than found alone. It was identified as a new element in 1781, and was first isolated as a metal in 1783. Its important ores include wolframite and scheelite. The free element is remarkable for its robustness, especially the fact that it has the second highest melting point of all metals (3,400 °C after Rhenium at 3,453 °C), and the highest tensile strength, which is of very high density and with a hardness close to diamond.

Tungsten, with essential applications in industry, aerospace and military, is a strategic commodity. The world market is dominated by China, which accounts for 80% of the world's mine production, giving rise to concerns over the security of supply of tungsten concentrates to western processors and industry end-users. This, along with other factors, have resulted in the EU categorising tungsten as a "Critical Raw Material," and the British Geological Survey ranking tungsten top of its metals "Risk List."

During gold exploration and mining activities in reefs and placers near De Jong-Zuid station (lat-long 5°04' and 55°13') at km 105,5 along the old railway line to Sara Creek (Lawaspoorweg or Landsspoorweg), miners found so-called blakaston (black rock) at Pakira Hill. The material had a high specific density, and was considered to be a valuable indicator of the presence of gold (Brinck 1956). This mineral is called ferberite, and contains wolfram. It is the iron-rich variety of wolframite and contains 76.3% WO_3. The mineral found at Pakira hill is a pseudomorph of scheelite (Brinck 1956). Pakira Hill lies within the Rosebel Gold Mine concession, where recent research has demonstrated the presence of scheelite in drill cores (Kroonenberg, personal communication).

Scheelite is also found at Omai in Guyana (Voicu 2000). Scheelite's formula is $CaWO_4$ and contains 80.5% WO_3. It is frequently present as an accessory mineral in Precambrian to Tertiary gold deposits (Voicu 2000).

While researching gold at Pakira Hill, Brinck found two other unknown minerals containing wolfram, which could not be identified at the time. A drill core contained a brown-black, possible pseudo-hexagonal mineral, which bears a lot of

Fig. 9.6 A sample of ferberite with orange ferritungstite (Naturalis Biodiversity Center, Leiden)

similarities with tungstenite FS_2 (Communication Kriegsman, Naturalis). Samples of orange-coloured contaminations of ferberite were analysed, using spectrographic and microchemical techniques, which were assumed to indicate the presence of ferritungstite. A sample is shown in Fig. 9.6 below.

Thanks to their high density, wolfram ores can be processed with a simple gravity separation process. The ore is commonly shipped in mtu's or metric tonne units, which are units of 10 kg or 1% of a metric tonne. Tungsten prices are generally quoted as US dollars per MTU of tungsten trioxide (WO_3). Theoretically, pure wolframite concentrate can contain 79.3% tungsten metal, but in practice, the grade of concentrate products acceptable for sale ranges from about 62% WO_3 to about 72% WO_3. Prices average around 350 $/mtu.

The average grade of the top-20 producing mines around the world is: 0.3% WO3. At this grade, it takes a mine production of 560,000 tonne per annum for a saleable production of 1,000 t or 100,000 mtu. Such an operation would require an investment on the order of 40 M $ in plant and equipment, with a capacity to process 100–150 tonne per hour, at a recovery of 60% with a reserve of 10 years' production.

Given the required scale of operation, it is worthwhile to investigate if Pakira Hill contains adequate resources of wolfram-bearing minerals. Road infrastructure of sufficient capacity is already available. It is worth noting that RGM is not pursuing exploration and mining activities on Pakira Hill due to its proximity to Nieuw Koffiekamp community. Alternatively, one can look for indications of scheelite associated with other known, accessible gold deposits.

References

Barakos G, Mischo H, Gutzmer J (2018) How potential mines can connect to the global REE market. Min Eng 70(8):30–37

Barakos G, Mischo H, Gutzmer J (2016) Strategic evaluations and mining process optimization towards a strong global REE supply chain. J Sustain Min 15(1), 26–35

Bosma W, Ho Len Fat AG, Welter CC (1973) Minerals and mining in suriname. Mededelingen Geologisch Mijnbouwkundige Dienst Suriname 22:71–101

Brinck JW (1956) Goudafzettingen in Suriname (Gold deposits in Suriname)–Thesis Leiden. Also published in 1956: Leidse Geologische Mededelingen 21, p. 1–246

Craig-David JR, Vaughan DJ, Skinner BJ, (1996) Resources of the earth: origin, use and environmental impact. Prentice Hall, pp 472

Dahlberg EH (1982) Geochemical investigation of magnetic and electromagnetic anomalies in the Upper Nickerie copper-rare earth mineralization area, Suriname p. 95–109. In: Laming DJC, Gibbs AK (eds) Mineral exploration techniques in tropical rain forest–AGID Report 7

Dahlberg EH (1987) Copper and phosphate mineralization in the lower Proterozoic mobile belt of Bakhuis mountains, Upper Nickerie, Western Suriname, Guiana Shield. Geologie en Mijnbouw 66:151–164

Dahlberg EH (2005) Proterozoic phosphorite in the Bakhuis mountains, p. 159–162. In: Notholt AJG, Sheldon RP, Davidson DF (eds) Phosphate deposits of the world, Vol 2

De Vletter DR (1984) Economic geology and mineral potential of suriname. Mededelingen Geologisch Mijnbouwkundige Dienst Suriname 27:11–30

Montagne DG (1964) An interesting pegmatite deposit in northeastern Suriname. Geol Mijnbouw 43:360–374

Ong E (2017) Bloomberg intelligence–Global Metals Market Trends

Ronald E (2016) Niobium in a nutshell Rockstone Research–Mining Geology HQ

Sheldon RP (1982) Analysis of preliminary studies of the Bakhuis mountains apatite deposits. Geologisch Mijnbouwkundige Dienst Suriname: unpublished report

Van Eijk HTL (1961) Preliminary abstract concerning pegmatite investigations in northeastern Surinam. Geologisch Mijnbouwkundige Dienst Suriname Jaarboek 1956–1958, p. 87–90

Van Kallen (1956) Verslag van de werkzaamheden op het terrein van Chin A Sen aan de Lawa–unpublished report

Voicu G, Bardoux M, Stevenson R, Jebrak M (2000) Nd and Sr isotope study of hydrothermal scheelite and hst rocks at Omai, Guiana Shield: implications for ore fluid source and flow path during the formation of orogenic gold deposits. Mineralium Deposita 35:302–314

Voncken JHL (2016) The rare earth elements–an introduction. Springer Briefs in Earth Sciences, pp. 127

Wong ThE, de Vletter DR, Krook L, Zonneveld JIS, van Loon AJ (eds) (1998) The history of earth sciences in Suriname, 479 pp

World Bank Group (2017) The growing role of minerals and metals for a low carbon future–World Bank Publications

Chapter 10
Non-metallic Minerals

10.1 Diamonds

Diamond is the hardest, most dense form of carbon. For its formation, diamond requires pressures reached only at depths of 150 km or more, followed by transportation through kimberlites. Because they are very hard and resistant to weathering, they accumulate in placer deposits. Kimberlites have not been located in Suriname, and in the absence of kimberlites, several theories with regards to the source of diamonds have been produced.

In Ghana, the source is presumably basic and ultrabasic metamorphic rocks. The only known kimberlite on the Guyana Shield is the diamondiferous Guaniamo kimberlite in Venezuela. This deposit is being mined extensively (Kroonenberg et al. 2019).

Most diamonds are small, and contain many imperfections, which makes them unsuitable for use as gems. About 80% of diamonds find an industrial application. Diamonds are measured in carats, where 1 carat equals 200 mg.

10.1.1 Exploration Results

Around 1880, two Americans found 60 good-quality diamonds in the upper Berg en Dal area, and since then, several finds of small to very small, alluvial diamonds have been reported from areas as far apart as the coastal area in the north and the Sipaliwini Creek near the Brazilian border in the south. Most of the diamonds have been found within the drainage area of the Suriname River.

Findings of diamonds were reported at the northeastern part of the Nassau Mountains area, near the Conglomerate Creek and the Paramacca Creek (Headley 1913).

M. Keersemaker, *Suriname Revisited: Economic Potential of its Mineral Resources*, SpringerBriefs in Earth Sciences, https://doi.org/10.1007/978-3-030-40268-6_10

Exploration for diamonds also took place in two concessions south and northwest of the Tafelberg, which is covered with Roraima Formation sandstones and conglomerates. This Formation is an important source of alluvial diamonds in Venezuela, Guyana and northern Brazil (De Vletter 1984).

In the 1950s Van Kooten carried out a diamond exploration program in the Rosebel-Sabanpassie area, focusing on two areas within a 4 km wide by 24 km long zone in savannah terrain, from the known Rosebel gold mining area near the Suriname River towards Sabanpassie near the Saramacca River. This region has been worked for gold, as discussed before, and diamonds were found during gold exploration activities, which warranted efforts to search for the extent and source of diamond occurrences.

The best results came from the heads of the so-called Y- and W-Creeks off the Rosebel Creek. Based on washed volumes of 171 m^3 of gravel at W-Creek and 64 m3 at Y-Creek, the recovered grades were 0.13 and 0.1 ct/m^3 respectively (Van Kooten 1954).

A quantity of recovered diamonds weighing 25 carats were analysed by a diamond firm in Amsterdam, which concluded that 17.5 carat were gemstone quality and 7.5 carat were industry quality, and that the diamonds appeared to be very similar to those found in Guyana.

It should be noted that due to screen sizes used in the early days, diamonds smaller than 1 mm^2 were not captured and included in the above estimates. Follow-up research in 1971 suggests that a considerable number of smaller diamonds were lost (Schönberger 1975).

Van Kooten stated that the likely source rock is a metamorphic subgrauwacke conglomerate, which forms part of a folded series of metamorphic sedimentary rocks. These rocks extend over a wider area than he explored. Schönberger analysed his results, and followed up with his own program of washing samples from pits and banka drill.

He concluded that the diamonds in the Rosebel-Sabanpassie area were derived from a local source in an area of 1 km^2 at the location where Van Kooten found his best-quality diamonds. He concentrated his efforts on finding kimberlite bodies in this area, though without success (Schönberger 1975). His theory is that kimberlite bodies are related to dolerite intrusions, which are present throughout the country. In reaction to a publication by Reid (1974), who proposes diamond-bearing kimberlites in West-Africa as the possible source for alluvial diamonds found in association with the Rosebel Formation in Suriname, Schönberger and de Roever (1974) concluded that diamonds should have a source in ultramafic rocks; but it is still a question which rock type is the source rock.

Also, recent IAMGOLD drill cores in the Rosebel Gold Mines concession, notably in the Koemboe area, showed the presence of volcaniclastic ultramafic rocks. These rocks are characterised by high contents of chromium and nickel (>1000 ppm Cr and >600 ppm Ni, respectively). Inspection of the Rosebel Gold Mines chemical database showed that rocks with such high Cr and Ni content are widespread throughout the whole area, suggesting that ultramafic volcaniclastic rocks have a larger distribution than just the Rosebel area (Naipal et al. 2019).

10.1.2 Economic Evaluation

In early-stage diamond exploration, values are presented as a diamond count, rather than a grade with a threshold, to have a minimum of one diamond per kilogram sample. As a general rule of thumb for economic potential, 1 carat/tonne of ore is viewed as high-grade (Stewart 2016); 0.2 ct/m^3 is suggested by Schönberger as order of size cut-off (Schönberger 1975).

The contained value per tonne can be extremely high, but it varies from deposit to deposit, depending on the quality and size distribution of the contained diamonds. Larger diamonds are much more valuable than smaller ones, and consequently, two diamond deposits with the same grade, which contain different proportions of large stones, will vary significantly in their value per tonne of ore.

Van Kooten's results are clearly below the threshold, but his indication that the source rock is found in a wider area–and Schönberger's analysis that the source rock may be far away– make it worthwhile to explore further. From a mining perspective, the savannah terrain with sandy material and low vegetation offers opportunities for low-cost mining. If the source is near Rosebel, accessibility of the location is not a major cost factor, although it must be remembered that RGM is the permit owner.

10.2 Uranium/Thorium

Uranium and thorium are radioactive materials which disintegrate spontaneously with the emission of energy, in the form of particles or rays. The breakdown of a large nucleus such as uranium into smaller nuclei is called nuclear fission. In nature, this decay takes place very slowly. This natural fission can be accelerated in the controlled environment of a reactor, to produce energy on a commercial scale. The energy from just one gram of ^{235}U is equivalent to 12 tonne crude oil (Craig-David et al. 1996).

In nature, uranium and thorium occur in various isotopes, of which ^{238}U and ^{232}T are the most common. The fissionable isotopes occur in much smaller quantities, so that uranium and thorium extracted in mining have to undergo a costly process to prepare the right reactor fuel.

10.2.1 Exploration Results

In 1959–60, an aeromagnetic survey, carried out over the Precambrian shield, showed a considerable number of radioactive anomalies. In 1971, a second survey, covering the northern part of the shield (in particular, the Avanavero area), was flown with a discriminating scintillometer, measuring uranium, thorium, potassium and the total gamma-flux. This survey indicated a few "good" and "fair" anomalies. A gamma-ray spectrometer ground survey in the Avanavero area showed a few minor anomalies, later confirmed by the aerial survey results. A similar airborne survey of the Rosebel Formation also revealed the occurrence of small anomalies.

On the grounds of the results of the first survey, and of radiometric checking of a number of drill cores, a consultant Sutton considered that there was a good chance of success in finding worthwhile uranium deposits. A programme for systematic follow-up by the GMD was started but was then discontinued. It was pointed out by consultants at the time that each Group or Formation could theoretically contain uranium deposits of one of the known types (de Vletter 1984).

Doeve (1966) mentions phosphuranylite (a uranium mineral) from the upper Kabalebo River area and huttonite and cheralite (both are thorium-bearing minerals) in sand samples from Jai Creek, Tapanahony River (also: Wong et al. 1998).

10.2.2   *Economic Evaluation*

Uranium has traded in a wide range, over the last 10 years, being negatively impacted by the shutdown of the nuclear industry in Japan. It is sold under long-term contracts with undisclosed prices, so it is difficult to know what would constitute a reliable reference price. Nowadays, a \$50 per pound rate is used as a long-term price, and a target grade for an open pit mine would be 2.2 lb/tonne; or, for exploration, approximately 0.1% would be an economic intercept over 100 m. For high-grade, underground deposits, a grade of 1% would be a significant intercept, over thicknesses of at least 2 m (Stewart 2016).

10.3   Kaolin

Kaolin, or china-clay, is the name for rocks which are rich in kaolinite. Kaolinite is a clay-mineral that can form the white high-purity material valued for porcelain making, as a filler in (glossy) paper manufacture, and in a wide range of other industries.

Large reserves of good quality kaolin exist under the bauxite deposits in the Moengo, and Onverdacht/Lelydorp areas near Paranam. In the latter areas, the thick overhurden is dumped on the kaolin in the mined-out portions of the mines, which is unavoidable in strip mining. However, in the Moengo mines, which have only very thin overburden, the kaolin in the mined-out areas is accessible. Table 10.1

Table 10.1 Chemical composition of sample material from Moengo

Al_2O_3	SiO_2	Fe_2O_3	LOI (1000 °C)	MnO_2	CaO	Melting point (°C)
40.1	44.3	1.55	14.08	<0.01	<0.05	1,400

below gives the chemical composition of arbitrary sample material from Bigi Gado in Moengo (Bosma et al. 1973).

Moengo Minerals started production in 2013 of meta-kaolin, which acts as an additive or substitute for cement in building materials.

The possibility of exploitation of the kaolin for filling and coating, in the paper industry, depends on whether its quality meets the rigid industry specifications of particle size (distribution), particle shape, brightness, grit content and rheology. The brightness is critical to the papermaker, with TiO_2 and Fe_2O_3 being the two most significant impurities (Bundy and Ishley 1990).

The best-quality kaolins mined in the world have TiO_2 and Fe_2O_3 grades in the range of 0.3–0.6% (Bundy and Ishley 1990; Raghavan et al. 1997).

10.4 Cinnabar

Mercury is a heavy metal with a silvery appearance. It is notable for being the only metal with a liquid structure under standard pressure and temperature conditions. It is also present in deposits worldwide, usually as mercuric sulfide or cinnabar. There are only a few countries where mercury is mined: notably Spain, China and Kyrgyzstan. Supply has been dwindling, while global consumption remains fairly stable at 2,000 tonne per year. As a result, prices have increased by a factor of 3–4 in the last decade.

10.4.1 Exploration Results

Cinnabar was first found by gold miners, who discovered scarlet red rocks in creeks in the Witlage area, between Toematoe and Tempati Creek near the Marowijne river. Samples were analysed, and appeared to contain high percentages of mercury.

Following up in 1915, Duyfjes found traces of cinnabar in ferritic material. The area was investigated by trenching, and 120 kgs of cinnabar-bearing material was recovered. The concentrate showed very low mercury contents (De Vletter 1984). It was concluded that the deposits did not have economic value, as the greater part of the resource had been removed by erosion, and the remaining tonnage was too small.

In Parbode (Feb issue, 2019), Kroonenberg describes the history of mercury exploration, including activities by Billiton in 1941 and 1953.

References

Bosma W, Ho Len Fat AG, Welter CC (1973) Minerals and mining in Suriname. Mededelingen Geologisch Mijnbouwkundige Dienst Suriname 22:71–101

Bundy WM, Ishley JN (1990) Kaolin in paper filling and coating. Appl Clay Sci 5 of 1991:397–420

Craig-David JR, Vaughan DJ, Skinner BJ (1996) Resources of the earth: origin, use and environmental impact. Prentice Hall, pp 472

Doeve G (1966) Delfstoffen in Suriname. Mededeling 15 van Geologisch Mijnbouwkundige Dienst van Suriname 94–107

De Vletter DR (1984) Economic geology and mineral potential of Suriname. Mededelingen Geologisch Mijnbouwkundige Dienst Suriname 27:11–30

Headley DE (1913) Diamonds in Dutch Guiana, engineering. Min J 95:888

Kroonenberg SB, Mason PRD, Kriegsman L, Wong TE, De Roever EWF (2019) Geology and mineral deposits of the Guiana Shield. Mededeling Geologisch Mijnbouwkundige Dienst Suriname 29:111–116

Naipal R, Kroonenberg SB, Mason PRD (2019) Ultramafic rocks of the Paleoproterozoic greenstone belt in the Guiana Shield of Suriname, and their mineral potential. Mededeling Geologisch Mijnbouwkundige Dienst Suriname 29:143–146

Raghavan P, Chandrasehar S, Damodaran AD (1997) Value addition of coating grade kaolins by the removal of ultrafine coloring impurities. Int J Miner Process 50: 307–316

Reid AR (1974) Proposed origin for Guianian diamonds. Geology 2(2):67–68

Schönberger H (1975) Diamond exploration between the Suriname and Saramacca Rivers (NE Suriname). Mededelingen Geologisch Mijnbouwkundige Dienst Suriname 23:228–238

Schönberger H, de Roever EWF (1974) Possible origin of diamonds in the Guiana Shield. Geology 2(10):474–475

Stewart A (2016) Making the grade—understanding exploration results. Resource World Magazine. https://www.resourceworld.com

Van Kooten C (1954) Eerste onderzoek op diamant: Rosebel—Sabanpassie. Mededelingen Geologisch Mijnbouwkundige Dienst Suriname 11:63 pp

Wong ThE, de Vletter DR, Krook L, Zonneveld JIS, van Loon AJ (eds) (1998) The history of earth sciences in Suriname, 479 pp

Chapter 11
Looking Ahead

Renewed interest in the Guiana Shield with Suriname has resulted in the South-American Exploration Initiative, or SAXI-program, which kicked off in 2018, with the overall aim to enhance the exploration potential of the region through an integrated program of research and data gathering. Exploration/mining companies are assisted in focusing their activities in areas of maximum prospectivity, while helping local Geological Surveys and universities (GMD and Anton de Kom University) in their role of providing pre-competitive data, information and competence.

The SAXI-program is managed by Amira from Australia, which is copying the format from a project it developed called WAXI, which focuses on West Africa. The core team remains involved in the project, and can build on experience from 12 years on Man-Leo Shield with many geological similarities to Guiana shield, partly involving the same group of researchers.

The Inter Guiana Geological Conference started in 1950, as a bi-annual event. After a hiatus of 43 years, the 11th Conference was held in Paramaribo in 2019, as evidence of the positive impact of this initiative on the region. The participation of some major mining companies as partners and sponsors of SAXI will ensure that positive results could end up as mine projects.

In the meanwhile, the discovery of a huge oil resource in the Stabroek block in Guyana's coastal waters, close to Nickerie (as pictured below), may offer cheap energy which could be used for refining and calcining Suriname's minerals, specifically from Bakhuis, which lies within close range (Fig. 11.1). The recent discovery of oil in Apache's exploration well in Suriname's own waters -announced in 2020- may result in an even better alternative.

Conditions seem to indicate a promising future, where some of Suriname's hidden potential could become unlocked, with the author's hope of successful and responsible development of its mineral treasures.

M. Keersemaker, *Suriname Revisited: Economic Potential of its Mineral Resources*, SpringerBriefs in Earth Sciences, https://doi.org/10.1007/978-3-030-40268-6_11

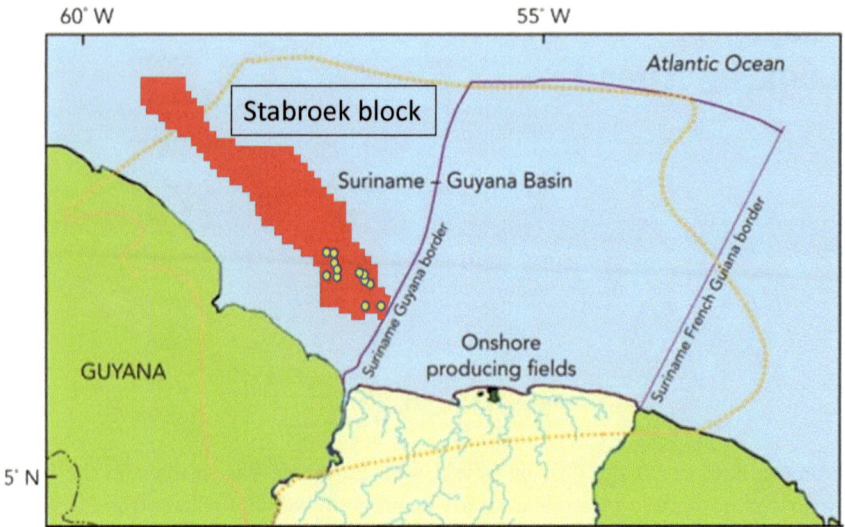

Fig. 11.1 Stabroek block with positive exploration wells offshore Guyana (adapted from https://www.exxonmobil.com and Oil & Gas Journal)